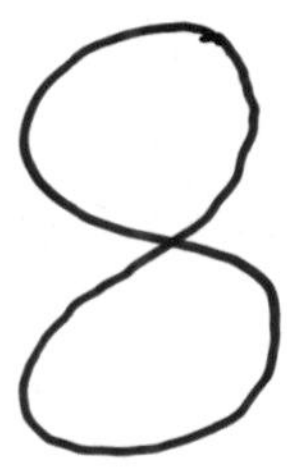

edexcel
advancing learning, changing lives

Jon Attwood
Barry Lambert
Peter Neal
Series editor: *Geoff Hancock*

Resistant Materials Technology

Edexcel A Level Design and Technology:
Product Design

Heinemann is an imprint of Pearson Education Limited, a company incorporated in England and Wales, having its registered office at Edinburgh Gate, Harlow, Essex, CM20 2JE.
Registered company number: 872828

www.heinemann.co.uk

Heinemann is a registered trademark of Pearson Education Limited.

First published 2008

12 11 10 09 08
10 9 8 7 6 5 4 3 2 1

British Library Cataloguing in Publication Data
A catalogue record for this book is available from the British Library.

ISBN 978 0 435757 78 6

Designed by Wooden Ark Studios
Typeset by Tek-Art
Original illustrations © Pearson Education Ltd
Cover design by Wooden Ark Studios
Cover photo/illustration © Pearson Education Lrd
Printed in Spain by Graficas Estella S.A.

Acknowledgements
The author and publisher would like to thank the following individuals and organisations for permission to reproduce photographs:
Pearson Education Ltd / Gareth Boden **pg 4** Pearson Education Ltd / Gareth Boden **pg 9** Pearson Education Ltd / Gareth Boden **pg 4** Pearson Education Ltd / Gareth Boden pg 6Alamy / ASP Norfolk Images **pg 31** Shutterstock / Andresr **pg 34** Alamy / David Askham **pg 41** Alamy / Helene Rogers **pg 42** Alamy / David J. Green **pg 45** Alamy / vario images GmbH & Co. KG **pg 55** Shutterstock / Glen Jones **pg 55** Shutterstock / pixelman **pg 57** Science Photo Library **pg 67** Techproducts **pg 67** BSI **pg 68** Pearson Education Ltd / Tudor Photography **pg 69** Shutterstock / Nigel Carse **pg 71** Shutterstock Stephen Finn **pg 71** Alamy / Itani still life **pg 71** Alamy / Caro **pg 71** Alamy / Itani still life **pg 71** Alamy / Photofusion Picture Library **pg 71** iStockphoto / Heinrich Volschenk **pg 71** Alamy / Inspirestock Inc **pg 71** iStockphoto / Daniel Loiselle **pg 71** Alamy / Mike Booth **pg 71** Evoluent **pg 73** Getty Images / Dan Krauss **pg 84** Science Photo Library **pg 87** Innocent Drinks **pg 92** Science Photo Library **pg 104** Science Photo Library **pg 106** Alamy / Corbis Premium RF **pg 106** Getty Images / Spencer Platt **pg 108** Getty Images / H. Armstrong Roberts / Retrofile **pg 110** Getty Images / Bill Kalis / Time & Life Pictures **pg 111** 1996 Forest Stewardship Council A.C. **pg 113** Nokia **pg 113** Audi UK **pg 114** akg-images **pg 117** Alamy / David Ball **pg 118** Bridgeman Art Library **pg 118** Alamy / eddie linssen **pg 119** Corbis / Peter Harholdt **pg 120** Bridgeman Art Library **pg 120** Zooid Pictures **pg 121** Alamy / Yadid Levy **pg 121** Alamy / imagebroker **pg 122** akg-images **pg 123** Alamy / dk **pg 124** Erno Goldfinger / www.aaschool.ac.uk **pg 125** Corbis / Fernando Bengoechea / Beateworks **pg 125** Harry Ransom Humanities Research Center The University of Texas at Austin **pg 126** Volkswagen Head Office UK **pg 126** Corbis / Bettmann **pg 127** Getty Images / Arnaud Chicurel / hemis.fr **pg 128** CARLTON, MEMPHIS 1981 design by ETTORE SOTTSASS / Thomas Dix / Vitra Design Museum **pg 129** Rex Features / Stephane **pg 130** Alamy / Redfx **pg 130** Bridgeman Art Library **pg 134** House for an Art Lover **pg 134** akg-images **pg 134** Alamy / dk **pg 134** Harry Ransom Humanities Research Center The University of Texas at Austin **pg 134** Alamy / Redfx **pg 134** Alessi / Hot Bertaa design Philippe Starck for Alessi **pg 133** iStockPhoto.com / Alexandra Draghici **pg 133** Alamy / Chris Willson **pg 136** Alamy / Richard Sheppard **pg 137** Alamy / Motoring Picture Library **pg 142** Sennheiser **pg 143** Alamy / Chris Laurens **pg 151** 1996 Forest Stewardship Council A.C. **pg 153** Alamy / Redfx **pg 158** Alamy / Flora Press **pg 158**

Author supplied images: pg 13, pg 15, pg 16, pg 17, pg 20, pg 23, pg 22, pg 162, pg 166, pg 167, pg 170, pg 171, pg 172, pg 173, pg 174, pg 175, pg 177, pg 179, pg 183 and pg 184.

Health and Safety Executive, The Manufacturing Institute, The British Automation and Robot Association, RotoVision, Ministry of Environment: Norway, United Nations

Every effort has been made to contact copyright holders of material reproduced in this book. Any omissions will be rectified in subsequent printings if notice is given to the publishers.

Websites
There are links to relevant websites in this book. In order to ensure that these are up to date, that they work, and that the sites are not inadvertently linked to sites that could be considered offensive, we have made the links available on the Heinemann website at www.heinemann.co.uk/hotlinks. When you access the site, the express code is 7786P.

Contents

Introduction

This book is designed to support Edexcel's GCE Resistant Materials specification. The book directly follows the structure of the specification and seeks to develop your knowledge, understanding, skills and application for designing products. Resistant Materials encompasses a wide range of design disciplines but is firmly rooted in the skills required to design and make high-quality products – products that are fit for purpose, satisfy wants and needs, enhance our day-to-day lives and, most importantly, give you the opportunity to demonstrate your design and technology capability.

This qualification emphasises two key factors: creativity and sustainability. We want you to explore ideas of originality and value, to question and challenge, to envisage what could be, but equally to achieve the results that will progress your career. This new qualification structure allows you to develop a range of skills and outcomes at AS level which will demonstrate your creativity and you will be able to apply these to a design-and-make project at A2.

All modern designers have to consider sustainable issues when designing new products. A sign of the modern technological age in which we live is that human actions have had a negative impact on our environment. New products should provide solutions rather than add to the existing problems of extractions and use of natural resources, pollution from manufacturing and disposal of large amounts of waste products.

Good design is vital to our world and economy; it is important, therefore, that you, as a future designer, develop a passion for designing your future.

How to use this book

This book is divided into four parts which correspond to the four units of the course: AS Units 1 and 2, and A2 Units 3 and 4. These sections will provide you with a depth of knowledge and understanding of Resistant Materials to help you study for examined units and to help you better understand the requirements of coursework units. All units follow the headings stated in the Edexcel specification content so that all the relevant information is covered in detail.

Summary of expectations for the unit	Unit content	Exam café
The first section of the unit summarises: • what you are required to do • what you will learn • how the unit is assessed.	This main section covers the subject content in depth. It: • explains what you will learn in each unit • helps you to understand the assessment requirements • provides further support and guidance throughout the unit.	The last section of examined units contains a revision section comprising: • a teacher area involving tips for studying • an 'ask the examiner' section involving worked examples of exam questions and practice exam questions.

***Table i** Structure of the book's AS/A2 units*

Support and guidance

A number of headings and boxes feature throughout the book designed to give you further support and guidance where appropriate.

General support and guidance common to all units.

- '**Getting Started!**' introduces each new section and asks some key questions relating to the content of each topic.

These help you to start thinking about how much you really know about a topic before you study it in depth.

- **'Factfile'** boxes contain information which may explain technical terms or further illustrate points in the main text.
- **'Weblinks'** provide suggestions for further study of key topics by giving relevant websites for you to research.
- **'Links to'** show you how information in one section can be cross-referenced to another section to help you fully understand a topic.

Support and guidance specific to examined Units 2 and 3.

- **'Think about this'** sections pose questions relating to the topic just covered. These are designed to test your knowledge and understanding of the topic you have just studied.
- There are a significant number of tables throughout examined units which will help you to revise significant points relating to many topics. In addition, the widespread use of figures helps to illustrate the main text.

Support and guidance specific to coursework Units 1 and 4.

- **'Activity'** boxes suggest relevant tasks you may want to undertake in order to develop your design and technology capability.
- **'To be successful you will'** boxes appear at the end of each coursework section enabling you to understand the assessment criteria and what you will need to evidence in order to gain as many marks as you can.
- Students' work is illustrated throughout coursework units to give you an idea of both the layout and content of each assessment criteria. They are examples that show good practice and should not be simply copied.

Exam café

An **'Exam café'** section features at the end of examined Units 2 and 3. This is an extremely useful section to use when revising for your exams and comprises two main sections.

1. Teacher area.

- A '**Revision summary**' gives you important information about studying for your exams.
- A **'Revision checklist'** provides you with a useful list of things to do when revising.
- **'Tips for answering questions'** gives you an invaluable guide to the command words used in each exam so you know what to expect from each question.

2. Ask the examiner.

- **'Worked examples'** for four questions are provided for each unit which show you what the examiner will be looking for in a range of different styles of question. The examiner will guide you through what each question is actually asking you to do and show you where marks will be awarded. High and low responses are used to illustrate each question.
- A number of **'Practice questions'** are given which directly relate to the content of each unit. These will prove extremely useful in testing your knowledge and understanding of a topic and provide invaluable exam practice.

How this course is structured

This is a four-unit course: two units at AS level and a further two units at A2 level which combine to make the full GCE qualification. The course will be assessed by both externally set examinations and internal assessment. For further guidance and a full outline of the content of Units 1–4 please refer to the **'Summary of expectations'** at the start of each unit in this book.

AS units

Unit 1: Portfolio of Creative Skills	Unit 2: Design and Technology in Practice
Internal assessment Internally set and marked by your school/college and externally moderated by Edexcel. **Number of marks:** 90	**External assessment** **Time:** 1-hour 30-minute examination set and marked by Edexcel. **Number of marks:** 70
You produce one portfolio that contains evidence of product investigation, product design and product manufacture. Photographic evidence must be supplied for the product(s) you have made.	You complete a question and answer booklet, consisting of short-answer and extended-writing type questions.
60% of AS course **30% of full GCE**	**40% of AS course** **20% of full GCE**

A2 units

Unit 3: Designing for the Future	Unit 4: Commercial Design
External assessment **Time:** 2-hour examination set and marked by Edexcel. **Number of marks:** 70	**Internal assessment** Internally set and marked by your school/college and externally moderated by Edexcel. **Number of marks:** 90
You complete a question and answer booklet, consisting of short-answer and extended-writing type questions.	You design and make a product. This is evidenced in your design folder with photographic evidence of you making the product and of the final product itself.
40% of A2 course **20% of full GCE**	**60% of A2 course** **30% of full GCE**

General exam advice

Preparation

'If you fail to prepare – prepare to fail' is probably the best piece of advice given to any student when it comes to taking exams. Therefore, it is important that you have developed a thorough knowledge and understanding of the topics covered and not just 'crammed' a few days before the exam. Useful preparation can include the following.

- **Organise your notes** as an organised file; this will save you hours when the time comes to revise. You could create:
  - Summary notes and "mind-maps" which outline the most important ideas of each topic in a visual manner instead of blocks of text.
  - Flashcards containing information that you need to have memorised. You could put topics on one side of the card, answers on the other. Flashcards will enable you to test your ability to not only recognise important information, but also your ability to retrieve information from memory.
  - Study checklists that identify the topics that you will be tested on for each exam. This checklist will enable you to break your studying into organised, manageable chunks, which should allow for a comprehensive revision plan with minimal anxiety.
- **Frequently read back over your notes after a lesson** to make sure that you understand the topic. Simply copying down notes from the board or from the textbook, without giving much thought to the content of your notes, will not be useful when you come to revise.
- **Frequently review your notes before the next lesson** and highlight any questions you may have. Do not be afraid of asking your teacher questions if you do not understand a topic.
- **Frequently test yourself on each topic** using the practice exam questions in the Exam café sections of this book and the 'Think about this' guidance throughout each unit.
- **Start to plan your revision early** so that you have more than enough time to cover all the topics in this exam and exams in your other subjects.

Revision and practice

Revision and practice are crucial to exam success. An exam not only tests your knowledge and understanding of topics but asks you to apply this to various contexts within each question. In order to build your confidence you should aim to:

- **Familiarise yourself with the format of the exam** using sample assessment materials and past papers so that there are no surprises on the actual exam day. Know how long you have and what sort of questions have been asked in the past.
- **Practise answering past questions** which provide you with invaluable experience of completing answers in a given time. Using exam conditions helps improve your planning and writing skills in producing focused responses.
- **Make the most of trial exams** as they help you to identify your strengths and weaknesses. Continue to revise topics that you are familiar with but do not avoid improving areas of weakness.

On the day

Everybody suffers from exam anxiety, which causes stress. However, there are some simple things that you can try which can manage your anxiety and reduce your stress levels.

- Prepare thoroughly and learn the topics well. Approach the exam with confidence and view it as an opportunity to show how much you have studied.

- Get a good night's sleep the night before and make sure you don't go to the exam on an empty stomach.
- Allow yourself plenty of time to get to school/college and even plan to get there a little early so you have time to relax before the exam.
- Don't attempt last minute cramming – you should feel confident that you have prepared.

During the exam

Now that you are actually in the exam hall and ready to start the exam there are a number of tips that can help you to perform to the best of your ability.

- **Preview the entire exam paper** by spending a short period reading through the paper carefully, marking key terms and deciding how to budget your time. Plan to do the easy questions first and the most difficult questions last.
- **Read each question carefully**, making sure that you know what the question is actually asking you and what is required in your response.
- **Plan your responses**, especially to extended writing type questions or essays. Use a blank piece of paper to jot down the key points and structure.
- **Write in a clear and legible manner** using black or blue ink so that the examiner can read your responses.
- **If you go blank then move onto the next question** and don't start getting stressed about the last one. You should have time to go back to that difficult question at the end of the exam.
- **Check your answers** and don't simply sit there staring into space if you have time to spare. You may spot a mistake or want to add more detail to a particular part – if you don't know the answer then have an 'educated guess'.

Examiners

All Edexcel examiners want you to do your best. None of them want to try to catch you out or fail you; indeed, examiners are sometimes briefed to give 'benefit of the doubt' to responses that are nearly correct. All examiners work from a mark scheme which is written by the principal examiner (who wrote the exam) and discussed in detail with examiners before they begin marking. A mark scheme indicates several acceptable responses to each question and also what is not acceptable. Therefore, examiners will be looking for 'triggers' in your work which relate directly back to their mark scheme.

It is extremely important to **read each question** thoroughly before you answer it. It is common for examiners to complain that too many students fail to answer the question that has actually been asked. Sometimes students try to write everything they know about the topic being examined instead of writing a focused response which can be allocated marks. It is also a common mistake to read the question too quickly and misunderstand what's being asked. For example, being asked for the *disadvantages* of a process rather than its *advantages*. It is often a very useful exercise to highlight the key words before you answer to make sure that you are giving the examiner the correct response.

Glossary of Terms

It would be extremely useful for you to have a copy of the specific unit content for the examined Units 2 and 3. These can be photocopied from your teacher's copy of the Edexcel specification or downloaded at www.edexcel.org.uk. Each section in the unit content carries a 'stem' explaining what you specifically need to learn for each examination.

Unit 2: Design and Technology in Practice

For example:
b) Polymers

Aesthetic, functional and mechanical properties, application and advantages/disadvantages of the following thermoplastics in the production of graphic products and commercial packaging:

(followed by the list of specific polymers)

Here, you need to be familiar with the specific properties of the polymers listed, where they are best used and why. The stem is further clarified by the use of polymers in graphic products (e.g. point-of-sale, product casings) and commercial packaging only. A question will not be asked, for example, on why PVC is best used for drainpipes as this is not a graphic product.

The following are the main terms used in Unit 2.

Key term in section stem	Meaning
Aesthetic properties	The visual qualities of materials.
Functional properties	The qualities a material must possess in order to be fit for purpose, e.g. the correct weight, grade, size.
Mechanical properties	The material's reaction to physical forces, e.g. strength, plasticity, ductility, hardness, brittleness, malleability.
Application	The quality of being usable for a particular purpose or in a special way; relevance.
Advantages/disadvantages	Qualities and features favourable to success or failure.
Processes	A description of the systematic series of actions needed to produce something.
Structural composition	How a material is made up.
Characteristics	Recognisable features that help to identify or differentiate one process from another.
Preparation	Action required before a process can begin.
Production/manufacture	The process of manufacture.
Concept	The general idea behind the use of quality assurance systems.
Principles	The distinct reasons for health and safety legislation.

Unit 3: Designing for the Future

For example:

Computer integrated manufacture (CIM)

Characteristics, processes, application, advantages/ disadvantages and the impact on employment of CIM systems to integrate the processing of production and business information with manufacturing operations, including:

(followed by a list of characteristics of CIM systems)

Here, you must study a wide range of aspects relating to the use of CIM systems. Firstly, you need to develop an in-depth knowledge and understanding of the features of CIM systems and how they are used to produce products. Then, you must explain the advantages and disadvantages of using CIM systems, in particular their effect on the modern workforce.

The following are the main terms used in Unit 3.

Key term in section stem	Meaning
Application	The quality of being usable for a particular purpose or in a special way; relevance.
Advantages/disadvantages	Qualities and features favourable to success or failure.
Processes	A description of the systematic series of actions needed to produce something.
Characteristics	Recognisable features that help to identify or differentiate one process from another.
Production	The process of manufacture.
Principles	The distinct reasons for something.
Impact	Effects felt as a result of man's intervention/modern systems.
Sources	Raw materials for processing.
Debate	Discussion involving opposing viewpoints.
Responsibilities	The duty and obligations of developed countries.

Unit 1:

Portfolio of Creative Skills

Summary of expectations

1 What to expect

In this unit you have the opportunity to develop your creative, technical and practical skills through a series of product investigation, design and manufacturing activities. The unit is divided into three different sections: Product investigation, Product design and Product manufacture. Each of these sections is separate and is not dependent on the other two, as opposed to one extended coursework project. All of the skills developed in this unit will be put to great use in the full design and make exercise in **Unit 4: Commercial Design at A2 level**.

This unit is set and marked by your teachers, then sent to Edexcel for moderation (sampling and checking of teacher's marks).

Product investigation

In this section you are free to choose any appropriate product(s) that interest you for your product investigation, so long as there is the opportunity to develop your skills in examining product performance, materials and components, product manufacture and quality issues. Alternatively, the choice of product(s) may be set by your teacher to ensure that a range of materials, techniques and processes are covered.

Product design

When working on this section of the unit you are not limited by the manufacturing or materials constraints of your school, or college, workshop. There is no requirement for your designs to be carried forward into a manufactured product. Therefore, you can design as openly as you like, developing creative and adventurous design, modelling and communication skills. Modelling during the development stage should be photographed to provide evidence. The individual design briefs/needs can either be set by yourself or given to you as a design exercise by your teacher.

Product manufacture

In this section you have the opportunity to develop many practical skills through making more than one product using a range of different materials. During product manufacture you do not need to design the product, as the focus is on gaining and developing practical abilities, as well as those of planning for production and prototype testing.

Manufacturing briefs should be set by your teacher to ensure that you can target specific practical skills and processes with a view to developing these into a broad set of skills and experiences in a variety of materials. Photographic records of all stages of manufacture are essential in providing evidence of advanced skills, level of difficulty and complexity so that you can gain the marks you deserve.

2 What is Resistant Materials Technology?

Resistant materials technology and product design, when combined correctly and appropriately, can be used to create a whole range of useful everyday objects.

Knowledge and understanding of materials, industrial and commercial practice, together with due regard for quality, health and safety and sustainable issues will help you through this course of study.

Through your work on this course, you will learn to communicate ideas and information, evaluate ideas, design, plan and make useful products. The work that you undertake in this course and the evidence that you are required to produce falls into **three** distinct sections: Product investigation, Product design and Product manufacture.

3 How will it be assessed?

Unit 1 is divided into three main sections with several sub-sections to focus your work. Each of these separate sub-sections contains assessment criteria that are allocated a certain amount of marks. A breakdown of each assessment criterion will be outlined in the *To be successful you will* sections of this textbook.

The maximum number of marks available for each section is 30, with an overall mark out of 90.

Section	Sub-section	Marks
Product investigation	A. Performance analysis	6
	B. Materials and components	9
	C. Manufacture	9
	D. Quality	6
Product design	E. Design and development	18
	F. Communicate	12
Product manufacture	G. Production plan	6
	H. Making	18
	I. Testing	6
	Total marks:	**90**

4 Building a portfolio

You will submit **one** portfolio that contains evidence for all **three** distinct sections. Your portfolio should contain a variety of different pieces of work that covers a wide range of skills and demonstrates an in-depth knowledge and understanding of Resistant Materials Technology.

Example: separate investigating, designing and making tasks:

Product investigation	Product design	Product manufacture
Product investigation 1: Analyse and research into an adjustable desk lamp.	**Design task 1:** (2D design) Design a mechanical device that can crush aluminium drinks cans to less than one-third their original length.	**Making task 1:** Manufacture a wooden jewellery box given working drawings and design requirements.
Product investigation 2: Analyse and research into a laminated dining chair.	**Design task 2:** (3D design) Design a portable artist's easel for use outside.	**Making task 2:** Accurately replicate an aluminium chess piece from an existing piece, using lathe and milling techniques.
	Design task 3: Design a range of coordinating jewellery based on a modular theme.	**Making task 3:** Manufacture the range of modular jewellery from Design task 3.

Example: combined design and making tasks:

Product investigation	Product design	Product manufacture
Product investigation 1: Analysis of an adjustable desk lamp.	**Design task 1:** Design an adjustable desk lamp with two axes of movement.	**Making task 1:** Manufacture an adjustable desk lamp with at least two axes of movement.
Product investigation 2: Analysis of a laminated dining chair.	**Design task 2:** Design a small piece of furniture that uses laminated forms in its construction.	**Making task 2:** Manufacture a small piece of furniture that uses laminated forms in its construction.

5 How much is it worth?

The portfolio of creative skills is worth 60 per cent of your AS qualification. If you go on to complete the whole course, then this unit accounts for 30 per cent of the overall full Advanced GCE.

Unit 1	Weighting
AS level	60%
Full GCE	30%

Product investigation (30 marks)

Getting started

In this section, you will analyse a range of existing commercial products using your knowledge and understanding of designing and making. You should take into consideration the intended function and performance of the product; the materials, components where appropriate and processes used during its manufacture; how it was produced and how its quality was assured.

FACTFILE:

When setting product investigation tasks you must take into account the following:

- Your chosen product must contain **more** than one material and process in order to access the full range of marks.
- You may investigate a range of different products over the course of your AS studies. However, for your portfolio, evidence of only **one** complete product investigation should be submitted. Evidence must **not** comprise the best aspects of a range of product investigations that you have undertaken.
- The submitted product can be chosen by you or by your teacher.

A. Performance analysis (6 marks)

When analysing your chosen product, you should determine what it was that the designer set out to achieve and then produce a technical specification that covers several key headings.

The technical specification should include the following:

- **Form** – why is the product shaped/styled as it is?
- **Function** – what is the purpose of the product?
- **User requirements** – what qualities make the product attractive to potential users?
- **Performance requirements** – what are the technical considerations that must be achieved within the product?
- **Material and component requirements** – how should materials and components perform within the product?
- **Scale of production and cost** – how does the design allow for scale of production and what are the considerations in determining cost?

Your specification points should contain more than a single piece of information, so that each statement is fully justified by giving a reason for the initial point. For example, it is not sufficient to say 'the material used is polystyrene'. You need to justify this point 'because it is tough and can be injection moulded'.

As part of this analysis, you should also look at one other existing similar product, using the same criteria identified in your technical specification. By finding out information on a similar product you can compare and contrast it with your own chosen product.

Performance analysis: workshop clamps			
PRODUCT 1: Quick action clamp	CRITERIA	PRODUCT 2: G clamp	ANALYSIS
This product has a specifically designed handle which allows the user to gain a firm grip in order to exert a large force.	Form	The tee-bar type handle allows a big mechanical advantage to be gained so that large forces can be exerted.	Both products have a common form in that they have a frame for a body. One end is moved towards the fixed end and pressure is exerted. Both are practical and easy to use.
The function of this product is for clamping quickly, but they are specifically designed to be used with one hand. They are mainly used by kitchen fitters where they are often holding a cabinet up with one hand and only have one free hand with which to use on a clamp. They are quick to use and to release.	Function	This type of clamp allows for a large force to be exerted. It uses a screw thread which creates a big mechanical advantage. This type of clamp is often used for holding pieces of wood together while the adhesive sets.	Both products essentially have the same basic function but their form means that they operate in slightly different ways.
The shape of the handle makes it easy to use with one hand since it uses a trigger type mechanism. The painted surface finish is smooth to hold and will not give rise to any injuries to the user.	User requirements	The t-shaped metal section on the clamp will withstand the forces which are trying to push them apart. The type of handle allows for a large mechanical advantage to be gained when doing the clamp up.	Although both clamps basically perform the same function they are very different in terms of how they are used and the speed with which they can be used and the forces exerted.
The quick release mechanism allows the moveable head to be pulled back very quickly. The trigger mechanism also allows the clamp to be closed up very quickly.	Performance requirements	The tough robust frame will withstand the large forces generated by the handle and the screw thread. It means that a large force can be applied to hold things together tightly whilst any adhesive sets. Since the clamp is all metal it can be used when holding pieces of metal together for welding.	The structural shape and form of the frame is critical since it stops it twisting, bending or deforming when subject to large forces.
The frame is formed from sheet steel in such a way as to improve its resistance to bending when a force is applied. The nylon cheeks that come into contact with the two pieces being held are relatively soft and will therefore not damage the pieces being clamped.	Materials and components requirements	The frame can be made from a number of materials such as cast iron, steel or in some cases aluminium. The threaded shaft is hardened and tempered so as to withstand the large forces involved.	Steel, although available in different grades and forms can be used for both clamps. Quite tough and hard it will withstand a good degree of wear and tear in the workshop.
The nylon cheeks are injection moulded in high volume and the springs would be bought in from a specialist spring manufacturer. Since the steel frame is easily press formed, cropped and spot welded, this type of clamp is quite cheap at £4.29.	Scale of production and costs	The frame is cast which even on a full-scale production run is quite an expensive operation due to the energy required to melt the metal. It is also quite a time consuming exercise in preparing the cope and drag and once cast secondary processing is required to clean up the frame and to remove the runners and risers. The heat treatment of the screw thread will also be quite expensive and therefore gives rise to a cost of £6.99 for a 150mm clamp.	The different processes and relative strength of the clamps give rise to a significant difference in the price.

Figure 1.1 *An example of the performance analysis for the comparison of two workshop clamps*

ACTIVITY:

Start your product investigation by carrying out a detailed physical study of your product. This will enable you to look at the product in closer detail and provide an opportunity to develop your communication skills.

1. Sketch a 3rd angle orthographic view of the product and, using Vernier callipers and/or a micrometer, accurately record the most important dimensions.
2. Construct an accurate 3rd angle orthographic drawing (to a suitable scale) of the product using a technical drawing board and equipment. Use British Standard dimensioning and labelling.
3. Draw the product in three dimensions using a pictorial drawing method such as isometric or two-point perspective. Use studio markers or coloured pencils to render the drawing to provide a realistic representation of the product.

Note: you could use suitable computer-aided design (CAD) software to perform activity tasks 2 and 3 in order to develop your information and communication technology (ICT) skills.

To be successful you will:

Assessment criteria: A. Performance analysis

Level of response	Mark range
Fully justify key technical specification points **(1 mark)** that relate to form, function, user requirements, performance requirements, materials and/or component requirements, scale of production and costs. **(1 mark)** Compare and contrast one other existing similar product using the technical specification. **(1 mark)**	4–6
Identify **(1 mark)** with some justification **(1 mark)** a range of realistic and relevant specification points that include references to form, function and user requirements. **(1 mark)**	1–3

Marks are awarded in the following order: you must achieve all the marks from the lower section first (1–3) before being awarded marks from the higher section (4–6). This applies to all the assessment criteria.

B. Materials and components (9 marks)

You will need to identify the materials and components used in your chosen product and apply your knowledge and understanding of their properties and qualities to suggest why in particular they have been selected for use. For example, Acrylonitrile butadiene styrene plastic (ABS) is used extensively in the electrical products industry for product casings, where specific properties are required. Product cases, such as a kettle body, must be suitable for high speed, high volume manufacture using the injection moulding process.

Advantages of using ABS include:

- it is an electrical insulator
- excellent surface finish can be achieved
- tough
- available in a wide range of colours
- relatively inexpensive to process.

LINKS TO:

Unit 2.1: Materials and components will provide you with the majority of information required to determine the choice of materials and components for a range of products.

As with many products, your chosen product could have been made effectively, in terms of quality and performance, from other materials and components. Therefore, you should investigate suitable alternative materials and components and, using advantages and disadvantages, compare them with the materials and components actually used. For example, wine racks can be made from a wide variety of materials, not just wood or metal.

ACTIVITY:

In groups of two or three, each select one different type of pencil sharpener. Each group has to carry out an analysis of the materials and components used for their selected type of pencil sharpener.

Still in groups of two or three, each person has to explain why their type of pencil sharpener is better than any other. This mini debate should uncover the advantages and disadvantages of the materials used and you should be able to arrive at a joint decision as to the most suitable material overall.

Sustainability is an important aspect to consider with any product and you should be able to explain the environmental effects of using the materials identified in the product in relation to one or more of the following.

- **Extraction and processing of raw materials** – what is the financial and environmental cost of using a particular material in terms of energy use and pollution? Can less material be used? Can recycled materials be used or can the product be designed so that it is easily recyclable?
- **Production processes** – do they require lots of energy or produce lots of waste products? Can the product be simplified to reduce the amount of production processes?
- **Disposal of products after their useful lifespan** – does the product minimise waste production? (Reduce, re-use, recover and recycle.)

LINKS TO:

Unit 3.4: Sustainability looks at these issues in greater detail. Although it is an A2 unit, the information given will be extremely helpful to this section.

Material and components analysis: G clamp			
Component	Why selected	Possible alternatives	Sustainability issues
Malleable cast iron	Malleable cast iron is relatively cheap since it only requires little adjustment to the composition of ordinary pig iron by the addition of carbon and silicon. It has very good strength under compression and it has good fluidity in the molten state which leads to the production of good castings. Cast iron also offers good corrosion resistance.	It is difficult to suggest any other material other than cast iron but on occasions aluminium can be used but it is softer and not so resistant to the bending forces. It can be alloyed to improve its mechanical properties such as toughness and hardness.	Cast iron is made by re-melting pig iron. At this stage the impurities are removed. Due to the temperatures involved a great deal of energy is used in this process but it does make use much recycled steel and iron and indeed the clamp once it reaches the end of its useful working life can be recycled.
Screw thread	The screw thread shaft is made from a low carbon steel or mild steel bar. It can be cold rolled to form the screw thread or it can be machined. It is relatively inexpensive material but it does need to be heat-treated once formed to improve its mechanical properties. In order to improve its resistance to corrosion the surface can be either zinc or chrome plated, or black finished, which involves heating it and dipping it into old oil.	There are very few alternatives indeed due to the forces involved. The material naturally needs to have a high torsion resistance due to the turning forces and pressures involved.	Much energy is used in the production of this part due to the fact that once the thread has been formed, the steel needs to be hardened and tempered and this involved a great deal of energy. Once the product reaches the end of its useful working life it can be recycled.

Figure 1.2. *An example of the analysis of the materials and components used in a G clamp*

To be successful you will:

Assessment criteria: B. Materials and components

Level of response	Mark range
Suggest, with reference to quality and performance, alternative materials and/or components that could have been used in the product. **(1 mark)** Evaluate, using advantages and disadvantages, the selection of the materials and/or components used. **(1 mark)** Describe the impact on the environment of using the materials and/or components identified. **(1 mark)**	7–9
Describe a range of useful properties that relate to the materials and/or components identified **(1 mark)** and justify their selection and use in the product. **(1 mark)** Identify alternative materials and/or components that could have been used in the product. **(1 mark)**	4–6
Identify a material or component used in the product. **(1 mark)** Describe a useful property of that material or component **(1 mark)** and justify its use. **(1 mark)**	1–3

C. Manufacture (9 marks)

You need to be able to identify, describe and justify the processes involved in the manufacture of your chosen product.

For example, the polymer casings for many electrical products are made using the injection moulding process. This is for many reasons, including its suitability for mass production, its speed and accuracy and the ability to create complex and interesting shapes the consumer may find appealing.

It is important to consider alternative methods of manufacture that could have been used in the manufacture of the product, so you should make clear **one** alternative method and compare and contrast it with the actual methods used. For example, the polymer casings for many mp3 players result in 'cheap' looking products. Some companies, however, use anodised aluminium or polished stainless steel casings that add 'weight' and have a higher quality 'feel' to them. Obviously, injection moulding cannot be used for metals so alternative processes such as extrusion and press forming need to be employed.

LINKS TO:

Unit 2.2: Industrial and commercial practice will provide you with many manufacturing processes used to produce a range of products.

ACTIVITY:

Study one of the many tools that you have available to you in the school, or college, workshop i.e. plane, hand drill, hammer, screwdriver etc. Determine what commercial manufacturing processes have been used to produce these items.

Now, think about alternative methods that could be used to produce those workshop tools 'in-house' by you and your classmates. How could you produce identical copies using the resources available to you, such as casting, turning, forming or fabricating?

Again, you should also consider, and describe, the impact on the environment of using processes identified in the production of the product. For example, the aim of many manufacturers is to reduce production costs by creating designs that use less material and less energy during manufacture and to reduce waste production. Controlling emissions of harmful substances during production is also a serious consideration, e.g. carbon dioxide produced from the burning of fossil fuels for energy.

LINKS TO:

Unit 3.4: Sustainability looks at these issues in greater detail. Although it is an A2 unit, the information given will be extremely helpful to this section.

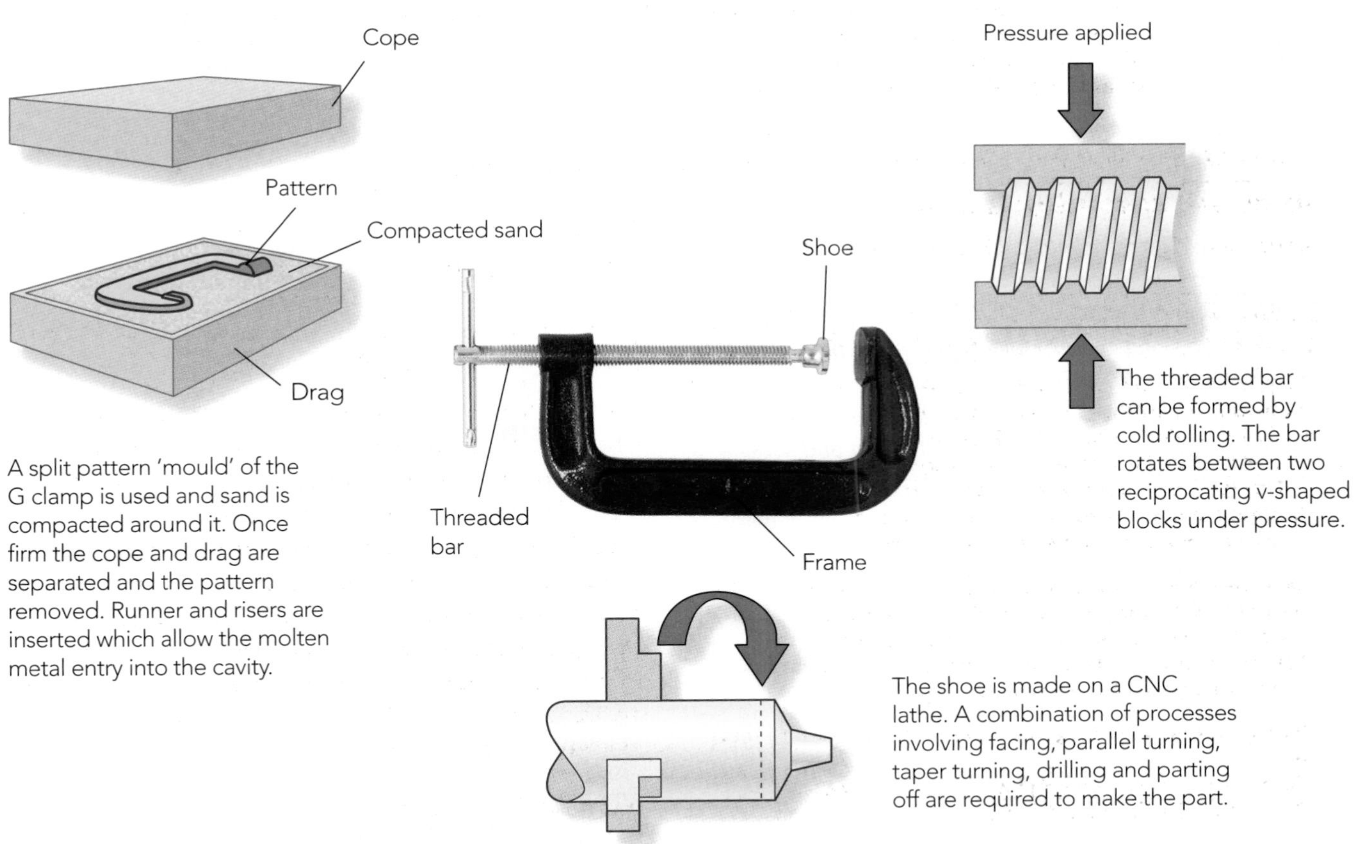

Cope
Pattern
Compacted sand
Drag
A split pattern 'mould' of the G clamp is used and sand is compacted around it. Once firm the cope and drag are separated and the pattern removed. Runner and risers are inserted which allow the molten metal entry into the cavity.
Threaded bar
Shoe
Frame
Pressure applied
The threaded bar can be formed by cold rolling. The bar rotates between two reciprocating v-shaped blocks under pressure.
The shoe is made on a CNC lathe. A combination of processes involving facing, parallel turning, taper turning, drilling and parting off are required to make the part.

***Figure 1.3** An example of the analysis of the manufacturing processes used for a G clamp*

To be successful you will:

Assessment criteria: C. Manufacture

Level of response	Mark range
Evaluate, using advantages and disadvantages, the selection of the manufacturing processes used in the product. **(1 mark)** Suggest one alternative method of production that could have been used in the manufacture of the product. **(1 mark)** Describe the impact on the environment of using the processes identified in the production of the product. **(1 mark)**	7–9
Describe: **(1 mark)** • a range of processes used in the manufacture of the product **(1 mark)** • and fully justify their use for the level of production of the product. **(1 mark)**	4–6
Identify, **(1 mark)** describe **(1 mark)** and justify the use of a manufacturing process used in the construction of the product. **(1 mark)**	1–3

D. Quality (6 marks)

All products will have gone through a series of checks and tests to ensure that in terms of quality and performance they reach the consumer in the best possible condition. You need to describe when and where quality control (QC) checks take place during the manufacture of your product, what the checks consist of and how they form part of a quality assurance (QA) system. For example, kettles are subjected to QC checks at various stages. The steel, or plastic bodies, are prepared, tested and examined for various defects including colour match, wear and surface finish. Electrical tests and checks are also carried out to make sure that it is safe to be used. Once all the different component parts are assembled and fixed in place, a final quality check is carried out to make sure it is all working safely and correctly.

You also need to identify and describe some of the main external standards that must be met during product manufacture and how they influence production and the final product. For example, what British Standards need to be adhered to when producing your product? These can cover a wide range of topics including materials selection, individual component testing, overall product testing or the standard of service and management. Some of these may result in the awarding of the British Standards Institute (BSI) 'Kitemark' or the European CE mark for QA.

FACTFILE:

Quality assurance (QA) is the system used by the manufacturer to monitor the quality of a product from its design and development stage, through its manufacture, to its end-use and the degree of customer satisfaction. In other words, QA is an assurance that the end product fulfils all of its requirements for quality.

Quality control (QC) is part of the achievement of QA. It involves the actual inspection and testing activities used by a manufacturer to ensure a high-quality product is produced.

External quality standards are used when testing, inspecting and verifying the overall quality of materials, components, products and systems. These **formal standards** are produced through standards organisations for national (British Standards (BS)), European (EN) or international (ISO) use.

LINKS TO:

Unit 2.3: Quality will provide you with the necessary information on issues relating to QC and QA procedures used in the production of a Resistant Materials Product.

To be successful you will:

Assessment criteria: D. Quality

Level of response	Mark range
Describe a range of QC checks used during the manufacture of the product **(1 mark)** and explain how the main relevant standards influenced the manufacture of the product. **(1 mark)** Describe a QA system for the product. **(1 mark)**	4–6
Identify, **(1 mark)** describe **(1 mark)** and justify the use of one QC check during the manufacture of the product. **(1 mark)**	1–3

Quality control of bought-in springs.

- The spring would be bought-in from specialist spring manufacturers. Engineers would monitor levels of quality at source and agree manufacturing specifications and tolerances suitable for the specific spring.

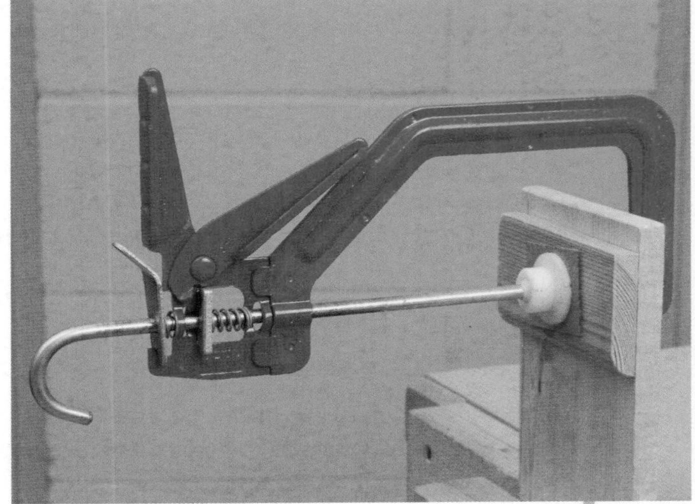

Quality control for injection-moulded nylon cheeks.

- Nylon pellets coming in from the factory would be checked against manufacturing specifications so that the correct grade and colour of material is being used.
- The injection mould would be visually inspected periodically for any damage or dirt/grit that would affect the moulding of the cheeks.
- Sampling and testing of batches of injection-moulded components would be carried out against agreed tolerances. Fine tuning of machinery would be required if sampling chart shows tolerances creeping.
- Sampling and testing of assembly of the parts would be carried out to ensure that they are fitting together correctly.

Quality assurance for quick action clamp.

Preparation

- Raw materials such as the nylon and mild steel would be sourced from reputable suppliers and quality control checks made on a regular basis.
- The springs would be outsourced to specialist manufacturers who would have to meet strict tolerances.

Processing

- QC checks feature heavily on the pressed steel shapes and the spot welding to make sure they are strong enough.
- QC checks on quality of painted surface finish.
- Sampling and testing of batches against agreed tolerances.

Assembly

- Sampling and testing of frames that have been spot welded.
- Correct assembly of spring mechanism and trigger.
- Cheeks fitting into steel frame.
- Final assembly test to determine fitness-for-purpose.

Finishing

- Packaged into plastic back and sealed. Packed into boxes and onto pallets – checking quantities for dispatch.

After-sales

- Guarantee assures customer that the clamp will be fit-for-purpose and will not break under normal conditions.

Figure 1.4 *An example of the analysis of quality issues related to a quick action clamp*

Product design (30 marks)

Getting started

In this section, you will have the opportunity to demonstrate your creativity and flair by using your design skills, through the production of a range of alternative ideas that explore different approaches to the same problem. Using the best aspects of your initial designs, you will develop and refine your ideas, with the aid of modelling, into a final workable design proposal that will satisfy a design brief or specific need.

Your designs do not have to be manufactured but the most viable products must be communicated to potential users. Any designer must sell their ideas by the use of presentation graphics or concept boards. Accurate working drawings and assembly drawings provide an audience with technical details of the product. Both forms of communication are invaluable in presenting an impression of the final product.

FACTFILE:

When setting product design tasks you must take into account the following:

- You can respond creatively and adventurously to one or more design briefs/needs.
- Design briefs/needs can be set either by you, or by your teacher, to produce solutions that are both fit-for-purpose and market viable.

E. Design and development (18 marks)

There are a number of possible starting points to this section. The design brief/need may be given to you by your teacher or you may define your own.

Two possible types of brief that you might want to use are:

- a focused design brief for a specific need or want
- a 'blue sky' project resulting in concepts using future technologies.

Design brief

Design a child's toy that helps him or her to learn and order numbers from one to nine. The toy must provide a storage facility for any loose pieces.

The toy must be stimulating for the child to use. A model can be used to test features such as scale, proportion and function and this can be done in 2D and or 3D computer simulations.

Design brief

Design a portable interactive 'shopping pod' which allows an electronic shopping list to be created at home and then taken to the supermarket.

Design specification

- The user should be able to access wi-fi in any store.
- The 'pod' must have a 'high-tech' appearance.
- The 'pod' should be no bigger than 150mm by 80mm.
- The 'pod' should be able to plug into an in-car charging unit.

Outcomes

- A series of high-quality design sheets which are well annotated and objectively evaluated against the criteria set out in the design brief for the need.
- A model of the 'pod'.
- A sample of the interactive user information screens.

You are not required to write a detailed design specification for each design task. However, the design brief must contain a range of design criteria that your final design proposal must meet. Therefore, you need to consider the design problem set and produce a range of alternative design ideas that focus on the whole, or parts, of the problem. It is not necessary for you to produce a wide range of alternative ideas. It is better to produce high-quality focused work than lots of lower-quality work.

Throughout your work you should explore different design approaches, applying your knowledge of materials, components, processes and techniques to produce realistic design proposals that satisfy the design brief/need. Design ideas should be objectively evaluated against the criteria set out in the brief/need, to ensure that your designs are realistic and viable. The use of detailed annotation is an important feature of design development and you should use it to explain details of design thinking and to offer thoughts on your design proposals.

It is important that you develop your own individual style when designing. Not everyone can produce beautifully presented and professional looking design sheets – the important thing is that you effectively communicate your design intentions. Experiment with a range of studio materials such as sketching with pencils, fine-liners etc., on different types of papers, for example a ballpoint pen on tracing paper or white pastel pencil on brown paper. Look at how professional designers present their design ideas and try to develop a more 'designerly' approach than your GCSE coursework projects.

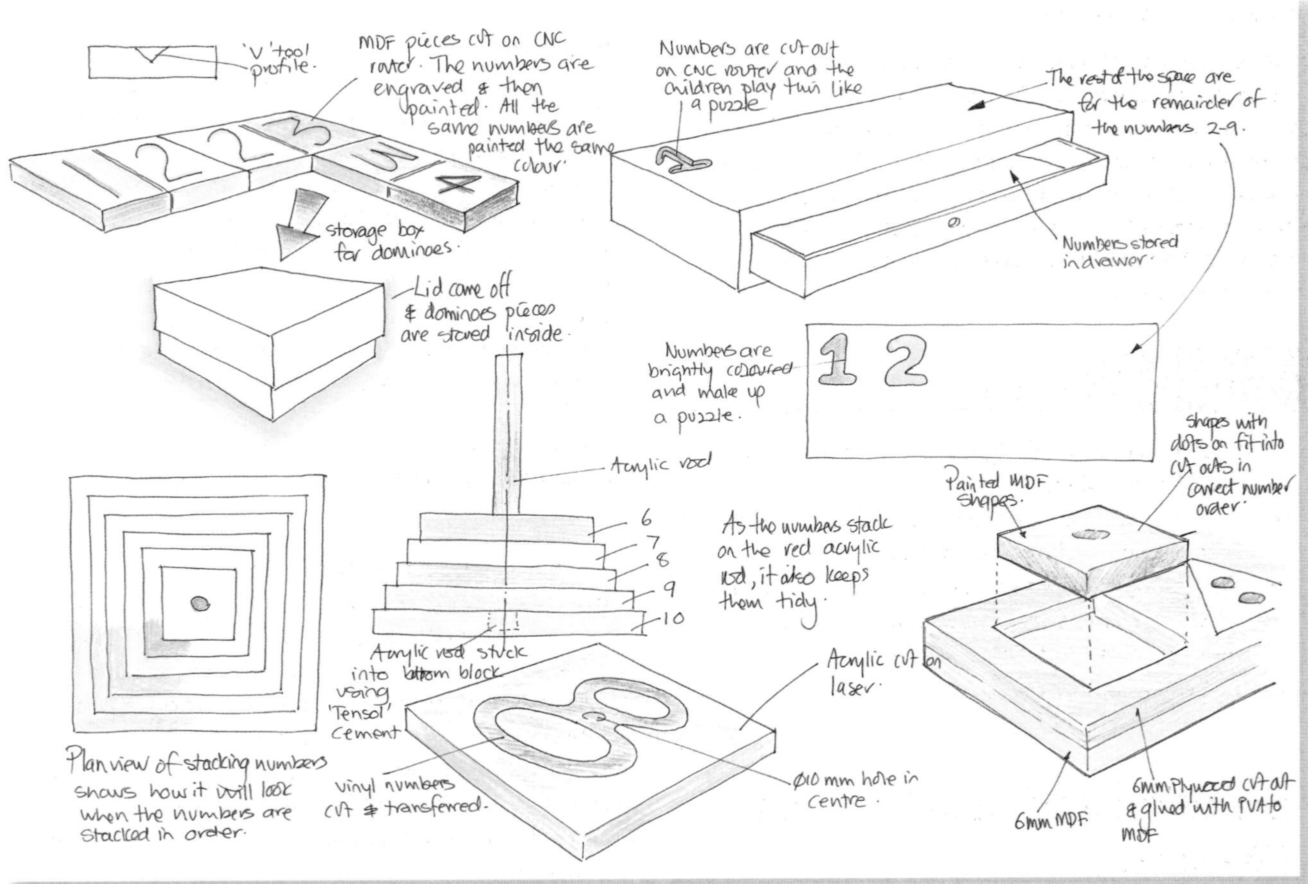

Figure 1.5 *This student's work for the design of a child's game is very structured and resembles GCSE work, although designs are communicated very well*

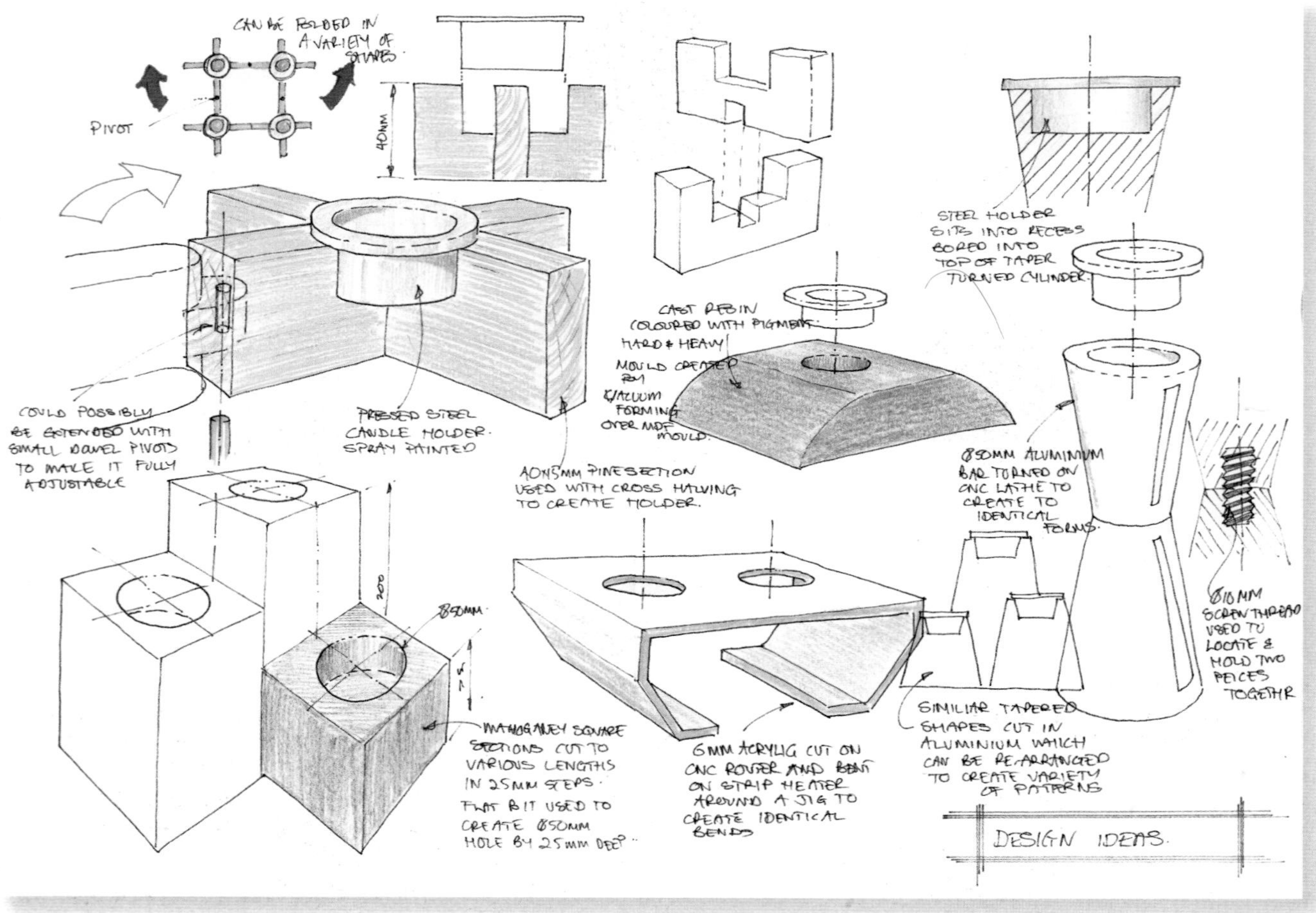

***Figure 1.6** This student's work for the design of a tea-light holder adopts a more professional approach*

ACTIVITY:

To practise your designing skills, set yourself a few small and manageable design tasks. These tasks should be focused and limited to three hours of design time. For example:

1. Design an in-car satellite navigation system that can be detached and used when walking. The product should be easily installed in a car and removed for use when walking, be easy to read and ergonomically sound to handle.
2. Design an outdoor garden lamp to be placed amongst the flower borders. It should be powered by a small solar cell.

When developing your initial design ideas, the design development cycle opposite can be used. Development is an important part of the design process and should be used to refine an initial idea into a workable design solution.

Modelling should be used to test features such as proportions, scale, function, sub-systems, etc. Modelling can be achieved through the use of traditional materials or 2D and/or 3D computer simulations. Evidence of 3D modelling should be presented using clear, well-annotated photographs. Card mock-ups should be photographed and included in your portfolio with associated evaluative comments clearly labelled.

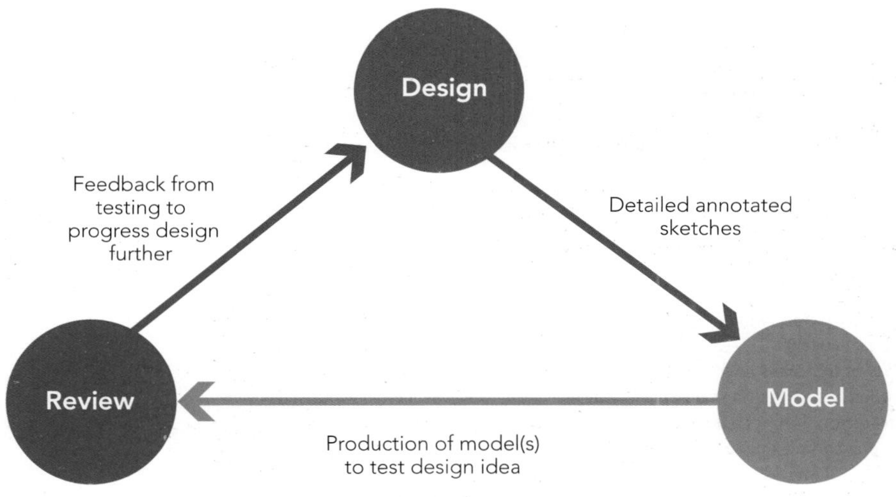

Figure 1.7 *Design development cycle*

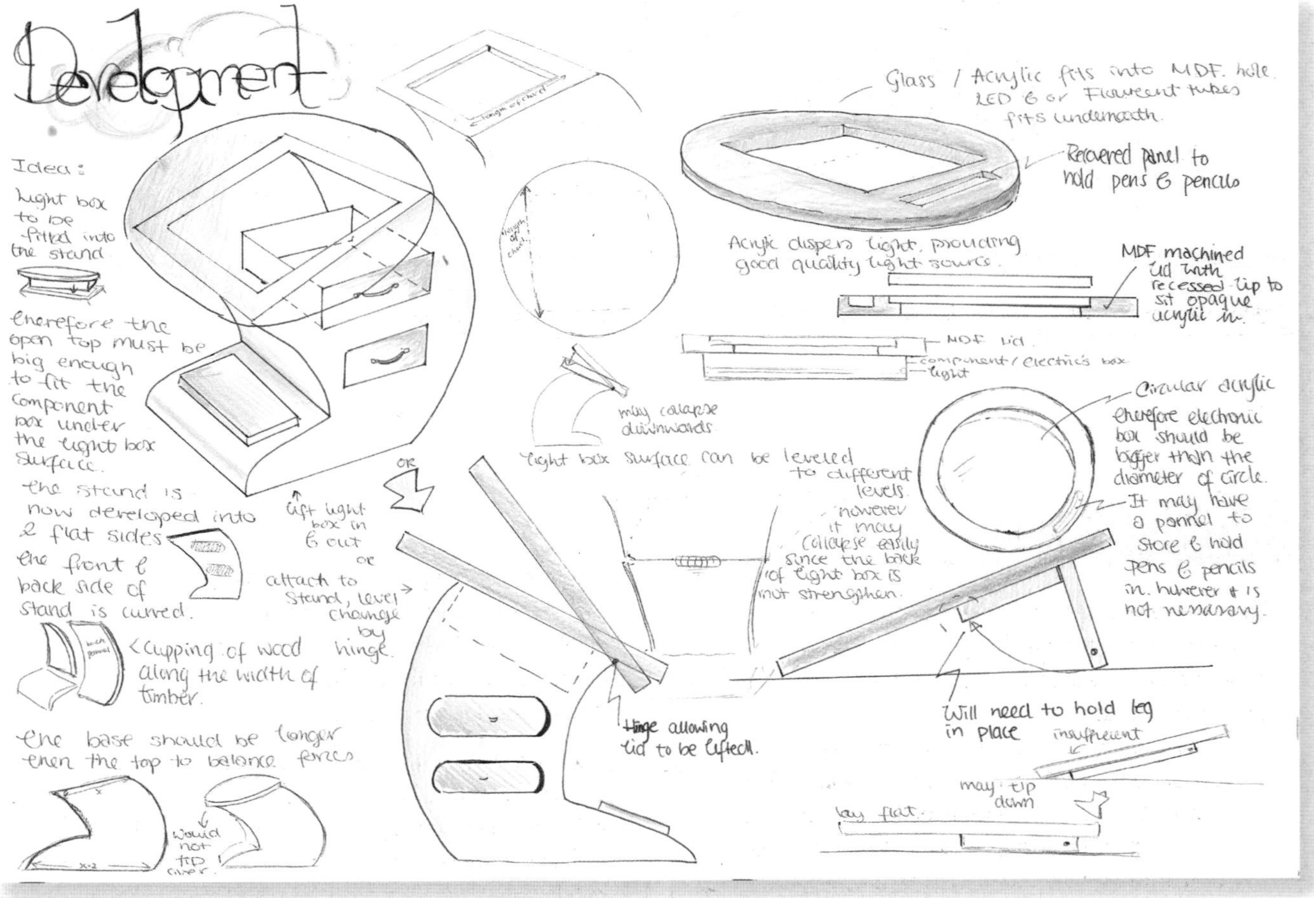

Figure 1.8 *A student's development of a desk which combines a light box suitable for tracing and paper storage to test both form and function*

To be successful you will:

Assessment criteria: E. Design and development

Level of response	Mark range
Present alternative ideas that are workable, realistic and detailed and that fully address the design criteria. **(1 mark)** Demonstrate detailed understanding of materials, processes and techniques in your ideas. **(1 mark)** Produce a final design proposal that is significantly different and improved compared with any previous alternative design ideas. **(1 mark)** Include technical details of materials and components, processes and techniques in the design proposal. **(1 mark)** Use modelling with traditional materials or 2D and/or 3D computer simulations to test important aspects of the final design proposal. **(1 mark)** Evaluate the final design proposal objectively against the design criteria in order to fully justify the design decisions taken. **(1 mark)**	13–18
Present realistic alternative design ideas. **(1 mark)** Present ideas that are detailed and address most design criteria. **(1 mark)** Use appropriate developments and use details from ideas to change, refine and improve the final design proposal. **(1 mark)** Present a final detailed design proposal. **(1 mark)** Use modelling to test some aspects of the final proposal against relevant design criteria. **(1 mark)** Objectively consider some aspects of the design brief/need in evaluative comments. **(1 mark)**	7–12
Present simplistic alternative design ideas. **(1 mark)** Present superficial ideas that address limited design criteria. **(1 mark)** Show developments that are minor and cosmetic. **(1 mark)** Present a basic final design proposal. **(1 mark)** Use basic modelling to test an aspect of the design proposal. **(1 mark)** Make evaluative comments that are subjective and superficial. **(1 mark)**	1–6

F. Communicate (12 marks)

When presenting your design and development work, it is essential that you communicate your ideas effectively. Evidence for this section can be found throughout the following areas:

(i) Through your design and development work

You should show evidence of 'design thinking' using any form of effective communication that you feel is appropriate. However, you should try to use a range of skills that may include freehand sketching in 2D and 3D, cut and paste techniques and the use of ICT. It is important to demonstrate a high degree of graphical skills, which will be shown through the accuracy and precision of your work.

The development of ICT skills is essential in the design and development of any product. Therefore, image manipulation software should be explored. When using CAD, you should ensure that it is used appropriately, rather than for show. For example, specialist CAD software to produce 3D rendered images is likely to be more appropriately used as part of development, or final presentation, rather than for initial ideas.

(ii) Through your presentation graphics and technical drawings

To effectively communicate final designs, a range of skills and drawing techniques should be demonstrated, which could include:

- **pictorial drawings** – isometric, planometric (axonometric), oblique and perspective drawings to convey a 3D representation of the product
- **working drawings** – 1st or 3rd angle orthographic, exploded assembly and sectional drawings to convey technical information
- **computer generated** – pictorial and working drawings, renderings, etc. using specialist software.

(iii) Through the quality of written communication

Annotation should be used to explain design details and convey technical information. You should make sure that all information is presented in a logical order that is easily understood. Specialist technical vocabulary should be used consistently with precision.

To be successful you will:

Assessment criteria: F. Communicate

Level of response	Mark range
Use a range of communication techniques and media including ICT and CAD **(1 mark)** with precision and accuracy **(1 mark)** to convey enough detailed and comprehensive information to enable third-party manufacture of the final design proposal. **(1 mark)** Use annotation that provides explanation and most technical details of materials and processes with justification. **(1 mark)**	9–12
Use a range of communication techniques, including ICT **(1 mark)** that are carried out with sufficient skill **(1 mark)** to convey an understanding of design and development intentions and construction details of the final design proposal. **(1 mark)** Use annotation that provides explanation and most technical details of materials and process selection. **(1 mark)**	5–8
Use a limited range of communication techniques **(1 mark)** carried out with enough skill **(1 mark)** to convey some understanding of design and development intentions. **(1 mark)** Use annotation that provides limited technical details of materials and processes. **(1 mark)**	1–4

Development

This is the work plan of the light box, diameter of full circle is 580mm and 420mm for the lighted part.

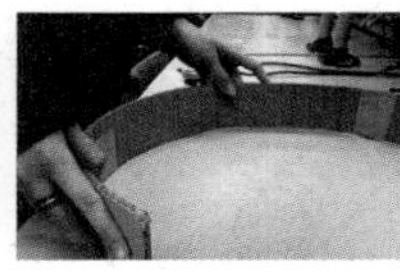

This is the base of light box, circular component box.

I've used the glue gun to stick the parts together.

The top should be placed on top of the component box.

The lit part should fit directly on top of the component box.

Some cardboard is cut off the side to fit light box.

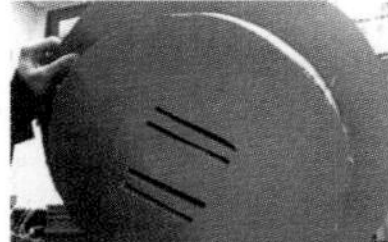

The bottom of the box should accommodate holes to release hot air generated.

The light box should be above knee height of client and be comfortable to draw on.

The light box should be slightly slanted for clients comfort.

Here is the table base of light box, made the same way with cardboard.

Light box fits on top of base.

The base top is too big for light box.

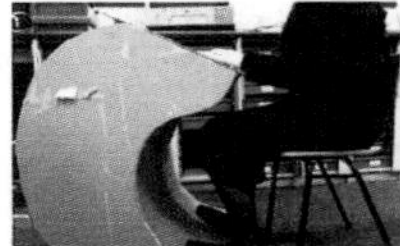

The curve of base should fit clients lower body.

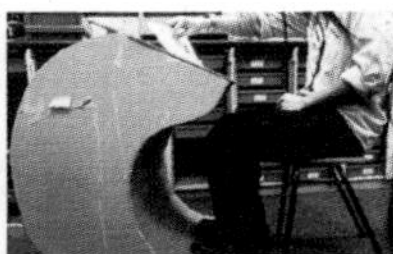

Different clients should also be comfortable when sitting.

The top of base is cut and glued again to fit base of light box.

The corner of base outside of light box.

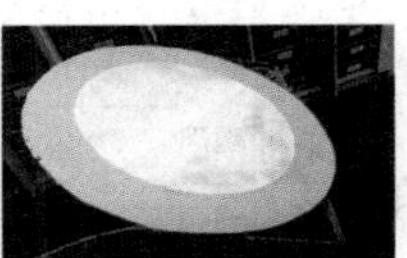

The box may either have a slanted side or the curve on the top made straight.

Client should draw more easily on light box.

Figure 1.9 *A student's modelling of a combined light box and drawing table*

Interactive information kiosk
Scale 1/12

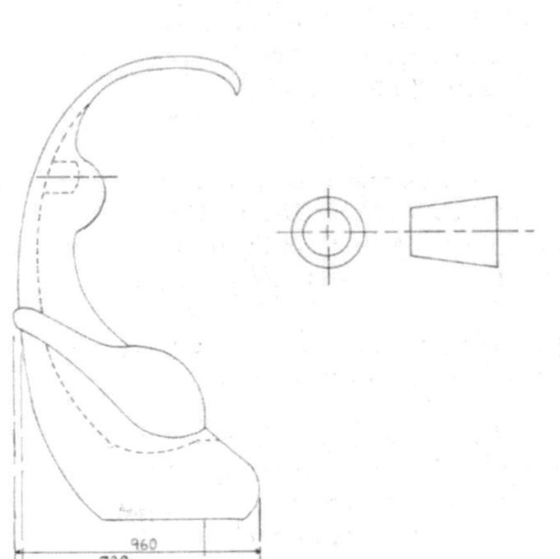

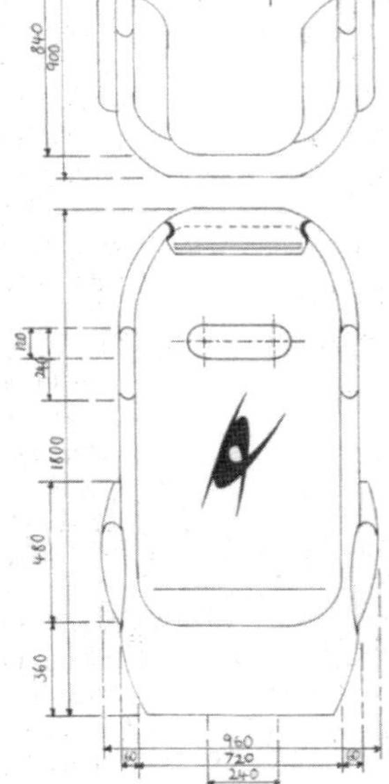

Figure 1.10 *A student's 3rd angle orthographic drawing of an interactive information kiosk proposal*

Product manufacture (30 marks)

Getting started

In this section, you will use your production planning skills and have the opportunity to develop your making skills through manufacturing **one or more** high-quality products to satisfy given design briefs/needs. You should use **a range** of materials, techniques and processes when manufacturing **a range** of products in order to build and develop a variety of skills and lay a foundation for more complex and challenging work in the future.

There are a number of potential starting points to product manufacture.

- Making a product previously designed in the design section (see pages 10–15). This takes the combined design and make task approach to your portfolio.
- Making a product from a detailed working drawing and manufacturing specification provided by your teacher. Here your teacher will specifically target skills and materials that you need to provide evidence for a wide range of techniques in your portfolio.

FACTFILE:

When setting product manufacture tasks you must take into account the following:

- You should produce one or more high-quality products that meet the requirements of the design briefs/needs.
- The design briefs/needs should contain requirements against which the final manufactured products can be measured.
- Some design briefs/needs may be set by your teacher to ensure a range of materials, techniques and processes are used.

Figure 1.11 *A student's part-made product*

Figure 1.12 A student's indoor light

G. Production plan (6 marks)

You need to produce a detailed production plan that explains the sequence of operations carried out during the manufacture of each product. A production plan should contain a work order or schedule, which could be presented in the form of a flow chart. The work order should include the order of assembly of parts or components and tools, equipment and processes to be used during manufacture.

Quality control points should also be identified throughout the production plan in order for you to produce a high-quality product. Specific quality checks should be described and not simply stated as 'quality control'. Information regarding important safety checks may also form part of detailed planning.

An important part of planning is the efficient use of time, so you should make sure that you consider realistic timings and deadlines. Where Gantt or time charts are used, you must make sure that they are detailed, cover all aspects of product manufacture and include achievable deadlines.

Consideration should be given to the scale of production of your products. Although you may be making one-off products, most products would be batch or mass produced, so you should consider the consequences of these scales in your planning and developing your awareness of commercial production.

Unit 3.2: Systems and control describes how flow charts are used to represent production processes.

To be successful you will:

Assessment criteria: G. Production plan

Level of response	Mark range
Produce a detailed production plan **(1 mark)** that considers stages of production in the correct sequence, **(1 mark)** with realistic time scales and deadlines for the scale of production. **(1 mark)**	4–6
Produce a limited production plan **(1 mark)** that considers the main stages of manufacture, **(1 mark)** with reference to time and scale of production. **(1 mark)**	1–3

Task		1 hour per week on practical task			
		Week 1	Week 2	Week 3	Week 4
1	Prepare materials	30 minutes			
2	Mark out and cut to length	30 minutes			
3	Mark out holes to be cut		30 minutes		
4	Rough out mortises		45 minutes	15 minutes	
5	Fine cut, clean up and fit mortises			45 minutes	
6	Rub down and seal				30 minutes drying time
7	Finish by applying top coat				30 minutes drying time

START
Set marking gauge to 42mm
Check measurements — No (return to previous step) / Yes
Mark parallel to square edge prepared on timber, by teacher, along the whole length
Set smoothing plane and plane timber down to the scored line in the timber
Check for correct width — No (return to previous step) / Yes
STOP

START
Set mortise gauge to 14mm and set to centre of timber
Check for centre — No (return to previous step) / Yes
Mark parallel to square, planed edge
Measure lengths along timber section and use try square and marking knife to mark across the grain at right angles to the square edge. Mark waste
Check marking out and dimensions against drawing — No (return to previous step) / Yes
STOP

START
Use mallet and firmer chisel to remove waste
Clean all edges up by paring with chisel
Check fit with other pieces — No (return to previous step) / Yes
Rub down and clean all edges and surfaces with glasspaper
Check for smoothness — No (return to previous step) / Yes
Apply sanding sealer and let dry
Rub down and apply a final top coat
STOP

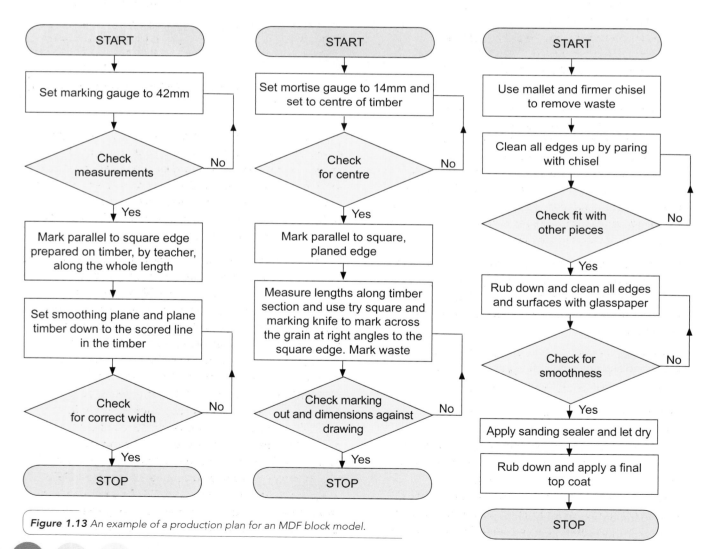

Figure 1.13 *An example of a production plan for an MDF block model.*

H. Making (18 marks)

You should produce one or more high-quality products that meet the requirements of the design briefs or needs that you have been given or developed yourself. The design brief/need must contain requirements against which the final manufactured product can be measured, so it is important when setting design requirements that they can be tested. Requirements may include dimensional parameters, finishes, etc., which are all objectively measurable requirements that can be tested for success.

Design brief

You are required to make a candelabrum as centrepiece for a dining table.

The candelabrum should:

- have a flat, stable base
- not weigh more than 1.5kg
- not damage the surface of the table that it will sit on
- be manufactured to a tolerance of $\pm$ 1mm
- be manufactured from flameproof materials.

Throughout your making activities, you should demonstrate your knowledge and understanding of a range of materials, techniques and processes by selecting and using those that are appropriate to the requirements of the task. You should consider properties and working characteristics of materials and the processes used to manipulate them. In addition to this you should also be able to justify your selections by giving reasons for your choices.

In order to develop high-quality skills you must apply your knowledge and understanding of a range of materials, techniques and processes. It is likely that you will produce more than one practical outcome during this unit.

Table 1.1 gives you some examples of possible projects and the wide variety of materials, techniques and processes open to you in your studies.

In order to achieve high marks in this section, you must show demanding and high-level making skills. Therefore, it is important that manufacturing tasks set by your teacher provide enough complexity and challenge to allow you to demonstrate your skill levels to the full. It is important to keep in mind, too, that the manufacturing tasks set in this unit should be designed to develop skills that you can call upon in your A2 coursework project. A single manufacturing project that involves a range of materials, processes and techniques that you can learn from can be as valid as two or three shorter but equally demanding exercises. However, by setting different exercises, the use of a range of materials, processes and techniques can be assured.

Project	Materials covered	Techniques and processes covered
Screwdriver	Body: aluminium Blade: silver steel	Body: turned on lathe, drilling, facing, parallel and taper turning; knurling for textured surface. Blade: forged end followed by heat treatment processes (hardening and tempering).
Ornamental flower trough	Mild steel	Mild steel: forging, welding, riveting; finished with 'Hammerite' plastic dip coating.
3D wooden puzzle	Mahogany	Marking out tools including try square, marking knife and mortise gauge. Smoothing plane, tenon saw, mallet and firmer chisels, glass paper and oil for finishing.
Bathroom tidy storage unit	Acrylic, ABS	Marking out tools, cutting, drilling, polishing and bending. Vacuum forming.

***Table 1.1** Examples of projects including materials, techniques and processes covered*

You will need to use a variety of skills and processes during your making activities, which may include computer-aided manufacture (CAM). Where this is a feature of your work, you should make sure that there is plenty of opportunity within the tasks to demonstrate other skills and competencies that you have gained through other making activities. While the use of CAM is to be encouraged, you must not over-use CAM. It is acceptable for you to dedicate one manufacturing exercise to the use of CAM in order to explore its capabilities, but you must offer evidence of other skills and techniques in other manufacturing exercises. Whenever CAM is used you must provide evidence of programming the computer numerical control (CNC) equipment. Where a mixture of CAM and other skills and techniques is used in a manufacturing exercise, CAM should not exceed 50 per cent of the work.

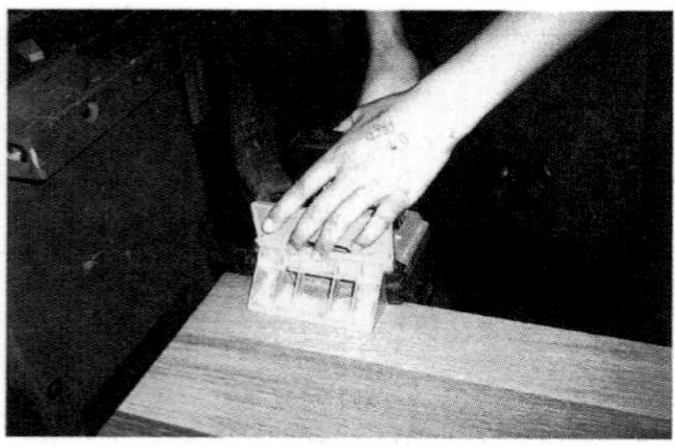

6. I used oak for the tabletop since the top needed to be hard-wearing and durable, because it is going to be used every day. Also, we had some other pieces of oak furniture in the room where the table was going to be used. When properly finished and varnished the grain and figure of oak is particularly attractive.

   I decided to use a biscuit jointer when joining the planks together to make the tabletop. I had to join planks because it is not possible to get a solid piece of timber the size I required. I used the biscuits as a form of locating and holding the planks together when gluing up. Sash clamps will be used with protective scrap wood between the tabletop and the cheeks of the clamps to ensure that no damage occurs to the tabletop edges when put under pressure.

7. I used a belt sander to sand down the top and to take out the minor steps where the planks join. This is an effective method because you are able to change the belts on the sander. I started with a very coarse belt to smooth the undulations in the surface. As the surface became flatter I was able to use a finer belt in order to make the scratches smaller. I completed the process with a palm sander with a very fine paper.

   I finished the table top with a hardwearing, heat-resistant varnish. Hot cups and plates will be placed on the table and I wanted to make sure they could not cause any damage.

8. Once all the legs and cross rails had been cut and dry assembled, I was ready to glue them together. I used PVA adhesive, a general-purpose woodworking adhesive. I was careful to remove any excess that came out of the joints when they were clamped, because it is very difficult to remove it once dry. I used mortise and tenon joints because they give a large gluing surface area and make for a very strong joint, which is what was required at the top of the table leg.

   I also made sure by measuring the diagonals that the pieces had gone together square. This was one of my major quality control checks.

Figure 1.14 *Justification of materials and processes as part of a student's detailed photographic evidence of a making activity*

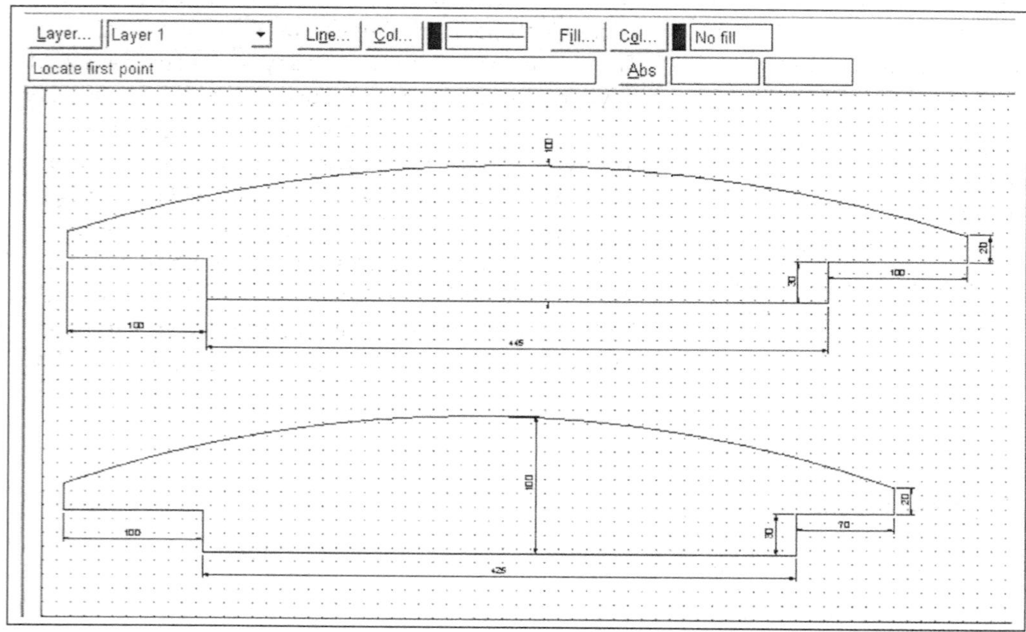

This is a more detailed view of the seat templates. I will be using these a great deal so therefore these need to be cut to perfection. The curves have been designed to give the utmost comfort for the client.

Figure 1.15 *CAM can be used to produce some components in a product. Here, the student has used CAM to produce multiple pieces for a chair*

Hazard	Risk	People at risk	Control measure
Abrasion – belt	Cuts/abrasions to hands	Teacher/ student	1. Sliding guard, where fitted to be 6mm above work. 2. Pupils queuing to use machine must stand behind yellow line. 3. Never leave machines running unattended. 4. Isolate machines after use. 5. Staff use only after recognised training. 6. Student use after training and at teacher's discretion.
Trapping – gap between belt and table	Trapping of fingers in gap between belt and table	Teacher	1. Adjustments to be made by competent staff. Post-16 students at teacher discretion. 2. Power to be isolated during repair and/or maintenance.
Projectile – dust and debris	Dust and debris being propelled into the eyes	Teacher/ student	Safety goggles provided in boxes next to machine.
Dust – respiratory irritant	Effects of prolonged dust inhalation	Teacher	1. Dust extraction system fitted and operational. 2. Dust extraction system serviced every year.
Dust – skin irritant	Effect of prolonged contact with wood dust	Teacher	Barrier cream available.

Table 1.2 *An example of a typical risk assessment for using a band facer in a school workshop*

Throughout your making, you should be aware of the risks involved in using specific tools, equipment and processes and should take appropriate precautions to minimise those risks. A risk assessment of all relevant equipment is an appropriate way of recording this awareness.

You will not be expected to produce a risk assessment for every piece of equipment you use. Annotated photographs of your making process should indicate where health and safety issues have arisen. For example, when using a band facer your photograph should clearly show you with your safety goggles on and with your hair tied back.

LINKS TO:

Unit 2.4: Health and safety describes the procedures for carrying out a risk assessment according to the Health and Safety Executive (HSE).

As proof of the quality of your making skills (and the level of demand of your work), photographs of your work must be evidenced to show that the product is complete, expertly made, well finished, etc. These photographs must clearly show any details of advanced skills, technical content, levels of difficulty and complexity of construction, so that you can achieve the marks you deserve. It is unlikely that a single photograph for each product will be enough to communicate all of the information required, so it will be better to take a series of photographs over a period of time during making. These should highlight the processes used and provide examples of precision and attention to detail that may not be otherwise noticed.

Realisation

Used router to remove excess waste on flexi ply.

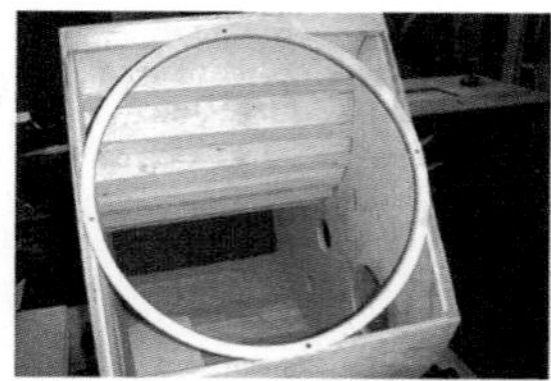
Fit circle rim into base and measure waste that has to be cut and draw line to emphasise.

Cut out waste on base using coping saw and sand down using sanding paper.

Glue consecutive rims together using PVA and tighten with G-clamps.

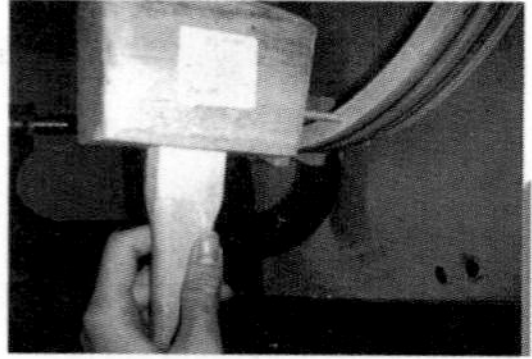
Dowelling, hammer wood sticks into holes of rims to secure them together.

Stick the bottom of light box to rim using PVA adhesive.

Used tool to level and remove extra acrylic. Smooth sides using wet and dry paper.

Used band saw to cut (teacher) and shape rim then sand down.

Sanded down rough and sharp edges.

I've used wet and dry paper to remove scratches on acrylic.

I've used the buffing wheel to polish the sides.

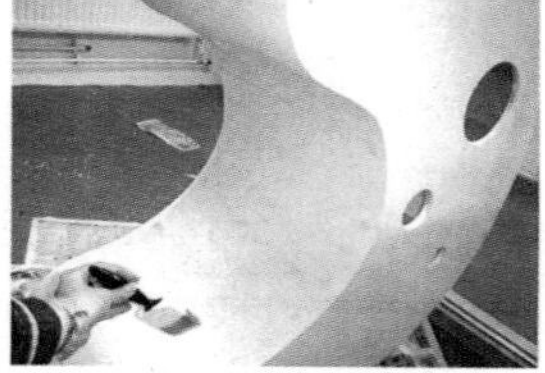
I've painted 2 layers of primer onto wood using a brush.

Figure 1.16 *Part of a student's detailed photographic evidence of the making process*

To be successful you will:

Assessment criteria: H. Making

Level of response	Mark range
Demonstrate a detailed understanding and justified selection of a range **(1 mark)** of appropriate materials **(1 mark)** and processes. **(1 mark)** Demonstrate demanding and high-quality making skills and techniques. **(1 mark)** Show accuracy and precision when working with a variety of materials, processes and techniques. **(1 mark)** Show high-level safety awareness that is evident throughout all aspects of manufacture. **(1 mark)**	13–18
Demonstrate a good understanding and selection of an appropriate range **(1 mark)** of materials **(1 mark)** and processes. **(1 mark)** Demonstrate competent making skills and techniques appropriate to a variety of materials and processes. **(1 mark)** Show attention to detail and some precision. **(1 mark)** Demonstrate an awareness of safe working practices for most specific skills and processes. **(1 mark)**	7–12
Demonstrate a limited understanding and selection of a narrow range **(1 mark)** of materials **(1 mark)** and processes. **(1 mark)** Use limited making skills and techniques. **(1 mark)** Demonstrate little attention to detail. **(1 mark)** Demonstrate an awareness of specific safe working practices during product manufacture. **(1 mark)**	1–6

I. Testing (6 marks)

After making each product, you should then carry out tests to check their fitness-for-purpose against the set design requirements. Your finished product, as far as possible, should be tested under realistic conditions to determine its success and to check its performance and quality. You should describe in detail any testing carried out and justify this by stating what aspects you are testing and why you are doing so. Tests should be carried out objectively, and it would be beneficial to involve potential users so that you can receive reliable and unbiased third-party feedback. Well-annotated photographic evidence is a very good tool to use when describing the testing process.

To be successful you will:

Assessment criteria: I. Testing

Level of response	Mark range
Describe and justify a range of tests carried out to check the performance or quality of the product(s). **(1 mark)** Objectively reference relevant, measurable points of the design brief(s)/need(s). **(1 mark)** Use third-party testing. **(1 mark)**	4–6
Carry out one or more simple tests to check the performance or quality of the final product(s). **(1 mark)** Reference superficially some points of the design brief(s)/ need(s). **(1 mark)** Record test results that are subjective. **(1 mark)**	1–3

Presentation of portfolio

Your portfolio of creative skills must be organised into **three** distinct sections clearly headed: Product investigation, Product design and Product manufacture. It is important that each individual piece of work provided for assessment is evidenced in the appropriate section. This will allow your teacher to easily mark your work and provide the Edexcel moderator with a clear indication of your skills and ability.

While there is no defined limit to the number of pages you should include, it is envisaged that all requirements of this unit can be achieved within 25–30 A3 size pages. You may choose to produce your product investigation in A3 or A4 format. You can also submit your work electronically for moderation provided it is saved in a format that can be easily opened and read on any computer system, i.e. a PDF document.

Authentication

It is extremely important that you sign the authentication statement in your Candidate Assessment Booklet (CAB) before your work is marked. If you do not authenticate your work Edexcel will give you zero credit for this unit.

Testing: third-party feedback

Joe was impressed with the way the bench looked. He thought it was a different, stylish design with a lovely finish that brought out the grain. The curves make it a unique style while it still looks traditional.

The bench can accommodate four people

Joe felt safe sitting on the bench. The sturdy structure, rounded edges and sanded surfaces made the bench safe for Joe to sit on.

The bench was very comfortable for Joe. He thought the bench would remain comfortable for a long time even though it doesn't look comfortable. The curve was well-researched with regard to the comfort of the person sitting.

Figure 1.17 *Third-party feedback was provided by a member of the target market group (TMG)*

Unit 2:

Design and Technology in Practice

Summary of expectations

1 What to expect

In this unit, you will develop your knowledge and understanding of a wide range of materials and processes used in the field of design and technology. It is important for you, as designers, to learn about materials and processes so that you can develop a greater understanding of how products can be designed and manufactured. You will also learn about industrial and commercial practices, the importance of quality checks and the health and safety issues that have to be considered at all times.

The knowledge and understanding that you develop in this unit can easily be applied to your Unit 1: Portfolio of creative skills.

2 How will it be assessed?

Your knowledge and understanding of topics in this unit will be externally assessed through a 1 hour 30 minute examination paper set and marked by Edexcel. The exam paper will be in the form of a question and answer booklet consisting of short-answer and extended-writing type questions.

The total number of marks for the paper is 70.

3 What will be assessed?

This unit is divided into four main sections, with each section outlining the specific knowledge and understanding that you need to learn:

2.1 Materials and components

- Materials
- Components

2.2 Industrial and commercial practice

- Scale of production
- Material processing and forming techniques
- Manufacturing techniques for mass production
- Joining techniques
- Material removal
- Heat treatment
- Conversion and seasoning
- Faults in woods
- Computer-aided design
- Modelling and prototyping
- Computer-aided manufacture

2.3 Quality

- Quality assurance systems and quality control in production
- Quality standards

2.4 Health and safety

- Health and Safety at Work Act (1974)

4 How to be successful in this unit

To be successful in this unit you will need to:

- have a clear understanding of the topics covered in this unit
- apply your knowledge and understanding to a given situation or context
- use specialist technical terminology where appropriate
- write clear and well-structured answers to the exam questions that target the amount of marks available.

5 How much is it worth?

This unit is worth 40 per cent of your AS qualification.

Unit 2	Weighting
AS level	40%

+ Finishe

Materials and components

Getting started

As a designer, you need to know about the properties of a wide range of materials and components so you can make informed choices about their use in certain products. What is the best material for a bike frame, for instance – aluminium or carbon fibre?

Would the sales of a product like a lawnmower increase if it were manufactured from a lightweight material such as aluminium rather than a plastic? Designers have to make these kinds of decisions with every product they design.

In this section you will need to develop knowledge and understanding, where appropriate, of the aesthetic, functional and mechanical properties of Resistant Materials Technology:

FACTFILE:

Property	Definition
Aesthetic properties	The sensory qualities of a material.
Functional properties	The qualities a material must possess in order to be fit for purpose, for example the correct weight, size, etc.
Mechanical properties	A material's reaction to physical forces, e.g. strength, plasticity, ductility, hardness, brittleness, malleability, etc.

Materials

Very often specific terms are used to describe particular properties of materials. A good understanding of those terms is necessary for a full understanding of the materials involved in Resistant Materials Technology.

Property	Definition
Plasticity	The ability of a material to return to its original shape once the deforming force has been removed.
Ductility	The ability of a material to be drawn or stretched.
Hardness	The ability of a material to withstand indentation, abrasion or scratching.
Brittleness	The tendency of a material to fracture under stress.
Malleability	The ability of a metal to be deformed by compression without being torn or cracked.

Table 2.1 *Properties of materials*

Metals

Metals can be divided up into three main categories:

- **Ferrous:** Metals that contain mainly ferrite or iron, such as steel which is made up of iron with various amounts of carbon. Occasionally ferrous metals have small amounts of other substances added to them to enhance their properties. Ferrous metals will rust and all are magnetic.
- **Non-ferrous:** Metals which contain no iron, such as aluminium, tin and copper. Non-ferrous metals do not rust, and they are not magnetic.
- **Alloys:** Metals that have been formed by mixing two or more metals and occasionally other elements in order to produce metals with enhanced properties. An example is stainless steel which is an alloy of steel and chromium. Adding chromium to the steel makes it less prone to rusting. Duralumin is an alloy of aluminium, copper and manganese. This makes the aluminium that is already lightweight into a material that is both lightweight and strong.

All metals originate from some kind of ore or mineral that is extracted from the earth. The production of metal requires some kind of heating or smelting process.

Ferrous metals

The ferrous metals that you will need to know about are cast iron, mild steel and carbon steel.

Cast iron

There are two types of cast iron, white cast iron and grey cast iron. White cast iron is not an easy material to work with as it is hard and brittle. Grey cast iron is the most useful. This material can be cast into intricate shapes and is often used for components on machines such as the tail stocks on lathes. It is a material that can be machined and when compared to steel it is relatively corrosion resistant.

One downside of cast iron is that it is brittle and will shatter, or break, if dropped and it is almost impossible to weld. It has high thermal conductivity which makes it ideal in situations where heat might be an issue. Examples of its use are in cookware and in braking systems in cars.

Mild steel

Steel is an alloy of iron and carbon. The more carbon that is added to the iron the harder the steel becomes. Mild steel has a carbon content of between 0.15 and 0.30 per cent. It is a very versatile material with many engineering applications. It is easily worked; it can be cut, machined, heated and hammered into shape. However, apart from case hardening, mild steel cannot be heat treated.

One major problem that steel does have is that it rusts. The iron in the steel reacts with oxygen in the air and oxidation begins to take place and this can be a problem. You only have to look at old cars to see evidence of oxidation in the form of rust. In order to stop oxidation, a barrier needs to be put between the steel and the atmosphere and the easiest way to do this is to paint the steel. This stops the oxygen in the atmosphere getting to the metal. Another method of preventing oxidation in steel is galvanisation, where a layer of zinc is put over the steel or plastic coating where a thermoplastic is melted over the steel.

Carbon steel

There are various types of carbon steel. There are medium carbon steels and high carbon steels. As more carbon is added to the steel the properties of the steel begin to change and their applications vary. Unlike mild steel, carbon steel can be heat treated to alter its properties. Medium carbon steels are used for things like garden tools. High carbon steels are used in products that need to be much harder like hammer heads, cutting tools and drills. Carbon steel is used in many engineering applications and is a very versatile material. It can be cut, shaped and machined.

The main disadvantage of carbon steel is that because it is a ferrous metal, it is prone to rusting. To prevent rusting, carbon steel should be painted or galvanised.

Name	Carbon content	Properties	Uses
White cast iron	1.7–2.9%	• Brittle and very hard • Cannot be machined	• Heavy machinery
Grey cast iron	2.5–4.0%	• Can be machined • Easily cast • Corrosion resistant	• Cast iron cookware • Disk brakes • Components for machines
Mild steel	0.15%–0.30%	• Ductile, tough, malleable • Has high tensile strength	• General engineering • Nuts and bolts
Medium carbon steel	0.30–0.70%	• Harder than mild steel but less ductile and less malleable	• Garden tools • Springs
High carbon steel	0.70–1.40%	• Properties can be improved by heat treating	• Hammer heads • Drills • Cutting tools

Table 2.2 *Properties and applications of ferrous metals*

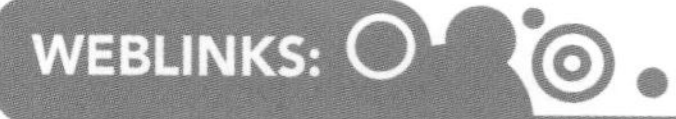

http://42explore.com/ironsteel.htm

This site has some excellent links to many iron and steel sites including production and manufacturing.

Non-ferrous metals

The non-ferrous metals that you will need to know about are aluminium, copper and zinc.

Aluminium

Aluminium is a pure metal. It is soft and malleable and conducts heat and electricity well. However, in its purest form, it is useless as an engineering material and is always alloyed to improve its properties. It is also extremely lightweight. For this reason it is a material that is extensively used in the aircraft industry. Aluminium will polish up and shine, but over a fairly short period of time the shine will disappear and a thin oxide layer will appear over the metal. This is an inert film of oxide and, unlike steel, where oxide in the form of rust flakes away and eventually destroys the material, the oxide film on aluminium actually protects the metal from further oxidation. For this reason, aluminium is often found on products that are used outside such as window frames.

One disadvantage of aluminium is that it is very difficult to weld. This is due to the very low melting point of aluminium and the oxide layer. As already mentioned, the oxide layer protects the metal but if an attempt is made to weld aluminium the oxide layer prevents the molten metal flowing and prevents a sound joint.

Copper

Copper is a malleable, ductile metal that has excellent heat and electrical conductivity. As it is ductile, it can be easily drawn into wire and because of this and its good conductive qualities is extensively used in the electrical and electronics industry. It is also corrosion resistant and for this reason is also to be found in household central heating and water systems.

Copper is a malleable metal. It can be worked cold, but it does work harden. As the copper is beaten, shaped or formed it becomes stiffer and harder to work. Eventually, cracks appear in the metal and it can break. When copper becomes work hardened it can be returned to its original malleable state by a process known as annealing. The copper is heated up and then allowed to cool. This returns the metal to its original state. Copper can be joined using both hard and soft soldering. It is a material that is often used for decorative items such as jugs or kettles. It can be easily cut and machined.

Unit 2.2 Industrial and commercial practice: Heat treatment.

Zinc

Zinc is easily worked and, like aluminium, has an oxide layer that protects it from oxidation. It is corrosion resistant and is perhaps best known for coating steel using the galvanisation process to protect steel from rusting. It is used in products such as buckets and water tanks and car bodies are dipped in zinc to help prevent corrosion. It is also used as an element in some soft solder fluxes. It has a relatively low melting point and is used in die casting.

Alloys

The alloys that you need to know about are stainless steel, duralumin and brass. Metals are alloyed to enhance properties – you might alloy a metal to make it harder or more resistant to corrosion or to improve machining characteristics.

Stainless steel

The problem with steel is that if it is left uncoated and subjected to the air, the oxygen in the air causes the metal to oxidise, or rust. One way of preventing this happening is to alloy the steel with another metal. Stainless steel is an alloy of steel, chromium and some nickel. The addition of these extra elements changes the nature of the steel and it becomes corrosion resistant.

Name	Melting point	Properties	Disadvantages	Uses
Aluminium	650°C	• Lightweight • Corrosion resistant • Good conductor	• Can crack under stress and requires constant annealing when worked • Does not withstand great loads	• Aircraft industry • Engine components • Castings
Copper	1100°C	• Good conductor of heat and electricity • Corrosion resistant	• Danger of electrolysis if joined to iron pipes in water systems • When worked requires constant annealing	• Electrical cables • Central heating • Printed circuits
Zinc	420°C	• Corrosion resistant	• When worked will become brittle	• Castings • Batteries • Galvanising

Table 2.3 *Properties and applications of non-ferrous metals*

+ Tin

Name	Composition	Advantages	Disadvantages	Applications
Stainless steel	• Chromium • Nickel • Steel	• Corrosion resistant	• Expensive • Hard to cut	• Kitchen utensils, pipes • Medical tools • Chemical and nuclear industries
Duralumin	• Aluminium • Copper • Manganese	• Lightweight • Strong	• Work hardens	• Aviation industry • Automobile industry
Brass	• Copper • Zinc	• Casts well • Easily to machine • Good conductor of heat and electricity	• Susceptible to cracking when cold worked • Needs to be constantly annealed	• Central heating valves • Electrical components • Ships' propellers • Plumbing fittings

Table 2.4 *Properties and applications of alloys*

Stainless steel has many applications including kitchen utensils, pipes, equipment for the chemical and nuclear industries and products for medical use. The disadvantage of stainless steel is that because it has chromium and nickel added to the steel, the final alloy is very hard and difficult to cut and may require specialist equipment.

Duralumin

Duralumin is an alloy of aluminium, copper and manganese. It is both very lightweight and at the same time very strong. It is a ductile and malleable material and has excellent machining characteristics. It is particularly useful in the aircraft industry where very strong, lightweight materials are essential. However, duralumin does age harden over time.

Brass

Brass is an alloy of copper and zinc that is corrosion resistant. It has excellent casting and machining characteristics. It is harder than copper, but can easily be joined using soft and hard soldering. It has good heat and electrical conductivity and is found in central heating systems and electrical plugs and sockets.

Polymers

The word polymer, in terms of Resistant Materials Technology, means plastics. There are two types of plastics: thermoplastics and thermosetting plastics. A thermoplastic is a material that, once produced, can be formed using heat, into a variety of shapes using different forming techniques such as injection moulding or blow moulding. Thermoplastics can be reheated, softened, shaped, reshaped and cooled many times over. They can be recovered and recycled easily.

Thermosetting plastics on the other hand are produced by mixing two components, a resin and a catalyst, together. These are usually in the form of liquids or pastes. Whilst they are apart they remain a liquid or paste but as soon as they are mixed together a chemical reaction occurs and the mixture becomes hard. Unlike thermoplastics, thermosetting plastics, once shaped cannot be heated and reshaped. Thermosetting plastics are therefore harder to recycle.

Thermoplastics

Thermoplastics are made up of long chains of molecules and these chains are held together by small cross links. The polymer chains are held together by mutual attraction called Van der Waals forces. When a thermoplastic is heated, the bonds between the chains of molecules weaken and become pliable. This means that the plastic can be shaped and formed. Once the heat is removed and the plastic cools down the chains reposition and the plastic becomes hard. This means that a thermoplastic can be heated and shaped many times.

Name	Properties	Disadvantages	Applications
Acrylic	• Stiff, hard, durable, can be easily scratched, good electrical insulator	• Brittle and can break • Scratches easily • Splinters easily	• Lighting • CD cases • Car lights • Baths
High density polyethylene (HDPE)	• Good electrical insulator • Chemical resistant • Impact resistant • Flexible	• Colour tends to fade over time • Can break under stress	• Water tanks • Water pipes • Buckets • Bowls
Low density polyethylene (LDPE)	• Good electrical insulator • Chemical resistant • Flexible	• Colour tends to fade over time • Can break under stress	• Washing up liquid bottles • Dustbin sacks • Cable insulation • Packaging film
Polyethylene terephthalate (PET)	• Good alcohol and oil barrier • Chemical resistant • High impact resistance • High tensile strength	• Can discolour • When used in containers for foodstuffs needs to be treated to prevent taste problems	• Fizzy drinks bottles
Polyvinyl chloride (PVC)	• Good chemical resistance • Weather resistant • Stiff, tough, hard, lightweight	• Can become brittle over time • Ultraviolet light causes brittleness	• Electrical wiring insulation • Pipes and guttering • Floor covering
Polypropylene (PP)	• Light, hard, impact resistant • Chemical resistant	• Ultraviolet light causes degradation • Oxidation can be a problem during manufacturing processes	• Medical syringes • Carpets • Kitchenware
Polystyrene (PS)	*Compressed:* • Light, hard, stiff, brittle, low impact strength	• Weak • Ignites easily	• CD cases • Refrigerator linings • Water tanks
	Expanded: • Buoyant, lightweight, good insulator	• Crumbles and breaks • Easily ignites	• Packaging • Insulation • Displays
Acrylonitrile butadiene styrene (ABS)	• Chemical resistant • Hard • Tough	• UV light causes degradation	• Mobile phones • Safety helmets

Table 2.5 *Properties and applications of thermoplastics*

Thermosetting plastics

Unlike thermoplastics, thermosetting plastics once formed cannot be reheated and changed. The molecules in thermosetting plastics link side-to-side and end-to-end. This linking is referred to as covalent bonding and the result of this bonding is a very rigid structure.

Name	Properties	Disadvantages	Applications
Epoxy resins	• Corrosion resistant • Electrical resistant • Good bond qualities	• Can cause allergic reactions • Suspected of health problems	• Adhesives • Paints and coatings • Electronics
Urea formaldehyde	• Strong, hard, brittle • Heat resistant	• Can emit toxic vapours during manufacturing process	• Electrical fittings • Domestic appliance components
Polyester resin	• Good electrical insulator • Heat resistant	• Brittle • Can crack	• Glass reinforced boats and cars • Garden furniture

Table 2.6 *Properties and applications of thermosetting plastics*

THINK ABOUT THIS!

Selecting the right material for the right job is very important. Different metals have many different characteristics. Think about a hand operated tin opener. What factors would a designer need to consider when considering the materials for a hand operated tin opener?

Woods

Hardwoods

Hardwoods are not necessarily physically hard. Balsa wood for instance is physically very soft but it is technically a hardwood. Hardwoods come from broad-leaved trees whose seeds are enclosed. Common hardwood trees include oak, mahogany and beech. Hardwood trees usually grow in warmer climates and take on average 80 to 100 years to reach maturity.

Because of the relatively long time they take to grow, hardwoods tend to be more expensive than softwoods. Very often hardwoods come from inaccessible places and this also adds to the cost.

Added to this, there is also an environmental issue. If a tree takes 100 years to mature it takes a long time to replace. For this reason there is some concern about the overharvesting of hardwoods. They produce tough, strong timber and often have a close grain.

Softwoods

Softwood is not necessarily physically soft. A plank of pine can feel as physically hard as a plank of oak! Softwood is the wood that comes from cone-bearing conifer trees. They have needles rather than leaves. Examples of softwoods are Scots pine (red deal), Parana pine and whitewood. You only need to know about pine in general terms.

Softwoods grow quicker than hardwoods. On average they take 30 years to mature and can be produced in plantations where they can be forested and replanted. As a result of careful planning, in particular by the Forestry Commission in the United Kingdom, softwood is a managed resource. As a consequence, softwoods tend to be cheaper to buy than hardwoods. As a rule, softwoods are easier to work than hardwoods.

Wood	Advantages	Disadvantages	Applications
Pine	• Straight grain • Easy to work	• Knots can make working difficult	• Construction • Roof joists • Floorboards • Furniture

Table 2.8 *Softwoods*

Composites

If two or more materials are combined together by some kind of bonding, a composite material is created. The resulting composite often has improved mechanical, functional and aesthetic properties when compared to the original materials from which it was made. Composite materials tend to have excellent strength-to-weight ratios and are useful for many projects.

Wood	Advantages	Disadvantages	Applications
Oak	• Strong, hard, tough • Works well • Durable	• Expensive • Heavy • Prone to splitting • Can be physically hard	• Garden furniture • Construction • High quality furniture
Mahogany	• Easy to work • Durable • Finishes well	• Grain can be variable • Prone to warping • Physical hardness varies	• Furniture • Veneers • Floorboards
Beech	• Physically hard • Tough • Polishes well	• Can be prone to warping • Not suitable for outside applications • Can be difficult to work	• Workshop benches • School desks • Furniture

Table 2.7 *Hardwoods*

+ Cedar, larch, redwood

+ Balsa + Jelutong

Glass reinforced plastics

Glass reinforced plastic (GRP) is a composite material that is manufactured from thermosetting plastic resin that has been reinforced using very fine glass. This glass comes in the form of either matting or in strips. Resin on its own is relatively strong when it comes to compressive strength but weak in tensile strength. By adding glass matting to the resin, the tensile strength of the final product can be considerably improved. In fact, glass reinforced plastic can be as strong as steel. The construction of GRP depends on thermosetting plastics that are produced by adding a catalyst to the resin. The resin remains a liquid until the catalyst or hardener is added to it.

There are advantages and disadvantages of GRP. It has good strength-to-weight ratios and is very resistant to corrosion. However, resin on its own is brittle and can shatter. It has been used as an alternative to steel in car bodies but has been found to be unsatisfactory if the vehicles were involved in accidents. It is used extensively in boat building but can be prone to osmosis. This is where water can seep into the material and cause the hull to delaminate.

Manufacturing a product in GRP	
Stage 1	A high-quality mould is prepared. This is the 'negative' of the product being produced.
Stage 2	A release agent is sprayed on to the mould. If this is not done, the final product will not be able to be removed from the mould.
Stage 3	A gel coat is applied to the mould. This is a thin layer of special resin that will form the outer skin of the final product. This will end up being a shiny layer and will be coloured. The colour is produced by adding a pigment to the gel coat.
Stage 4	At this point alternative layers of resin plus catalyst and glass matting are placed in the mould. It is important that the resin is worked well into the matting to ensure the best finish. Usually in an item like a fibreglass canoe, there should be about six layers of glass matting impregnated with resin.
Stage 5	Once the 'laying up' process has been completed, the work is left to cure over night and then it can be gently eased out of the mould.

Table 2.9 *Manufacturing a product in GRP*

When working in resin and glass matting safety can be an issue. The resin and catalyst can give off toxic fumes and must be mixed in a well-ventilated room. Gloves must be worn when using glass matting.

Figure 2.1 *Typical Norfolk Broads cruiser, manufactured from glass reinforced plastic.*

Carbon fibre

During the last few years, carbon fibre has become more common. This is produced commercially in a similar way to GRP. The carbon fibres in carbon fibre composites, which are able to take tensile loads, are set onto a polymer matrix that is able to take compressive loads. Carbon fibre constructions are much stronger than GRP and are ideal for high-performance structural applications such as aircraft, sports equipment and racing car manufacture. Carbon fibre composites have excellent properties and in some instances are able to replace more traditional materials.

THINK ABOUT THIS!

Carbon fibre is becoming a composite that is being used in many industries. Why do you think this is the case? Why would an aircraft manufacturer, for example, choose carbon fibre for their product over more traditional materials?

Medium density fibreboard

Medium density fibreboard (MDF) is one of the commonest and most widespread composite materials. It is made mainly made from wood waste. Sometimes it is manufactured from especially grown softwoods and this is an issue that is of concern to some environmentalists. MDF is manufactured from wood chips that are first subjected to heat and pressure. This is done to soften the fibres in the wood and produce a fine wood pulp. In order to produce a material that has a uniform structure, the wood pulp is then mixed with synthetic resins that are designed to bond the fibres of the wood pulp together. This is then heated and put under pressure to produce a sheet material that is uniform in thickness with a fine textured finish. After cooling the sheet material is trimmed.

MDF can be laminated in order to improve its strength or to improve its aesthetic qualities. It is a material that can be cut and machined like wood but has no grain so there is no danger of it splitting or warping. However, one issue that should be taken into account when working with MDF is that when it is cut or sanded, very fine fibres and dust are produced. Great care must be taken when drilling, cutting and especially sanding MDF. Dust extraction systems should be used and dust masks should be worn as the dust given off can cause irritation to the skin, throat and nasal passages.

MDF is used in products such as furniture or kitchen units or worktops. It is usually veneered with a thin layer of high-quality wood or Formica® to give it a good finish. Its greatest disadvantage is its weight – MDF can be very heavy.

Chipboard

Chipboard is another common composite material. In chipboard (like MDF) wood particles are glued together under heat and pressure and the result is a rigid board that has a relatively smooth surface. It comes in normal, medium and high-density form. The most common is normal density which is a fairly soft material. At the opposite end, high-density chipboard is solid and hard. Chipboard is only really suitable for internal applications and is often used for things like work surfaces in kitchens where it is veneered to give it an attractive finish, flat pack furniture and internal fire doors. If it gets wet, chipboard soon becomes waterlogged, starts to swell and breaks down.

Laminates

Sometimes using solid woods in the production of a piece of practical work is not economical or suitable. It may be better in some situations to use laminates. A laminate is a material that is produced by bonding together two or more layers of material. Laminates are materials that are usually very stable. They do not warp or misshape and in many instances are easier to work than solid woods. They come in sheet form that is easy to mark out and easy to cut and shape.

Although laminates are often much cheaper than solid wood, the aesthetics of the material may not be as good as solid wood. The surfaces of laminates are rarely good enough for a final finish. In many instances, they need to be finished either by veneering or even painting to make them look good. Laminates are often used in flat pack furniture. However, when used in such applications, because the material is relatively thin, the normal wood joining processes cannot be used. Laminate sheets cannot be joined using traditional wood joints such as dovetail or finger joints. In flat pack furniture, knockdown fittings are usually used.

Plywood

Plywood is manufactured from layers of veneers that are bonded together using glue. Veneers are thin sheets of wood usually about 1mm thick that have been obtained from timber using the rotary cut process. This is where a log is centred on a lathe and turned against a broad cut knife, producing thin slices of wood veneer.

Plywood is a very strong and stable sheet material because the layers of veneer are bonded in such a way that grain direction on each sheet is at 90° to the sheet of veneer above and below it. The number of layers is always an odd number – you will only find three, five, seven ply, etc. As the grain on each layer goes in different directions and because of the uneven number of layers, plywood on the whole is very stable – it does not warp or distort. However, if plywood becomes damp or wet, the layers can delaminate and the layers can come apart. Some plywood known as marine ply, using waterproof glues, are used in the manufacture of boats, but these tend to be more expensive than interior plywood. Plywood has many applications. It can be used for flooring as a constructional material and when applied with a decorative veneer it is used as a material in flat pack furniture. It is used extensively in school design and technology workshops as a material in projects.

Blockboard

Blockboard is made up of strips of softwood that are approximately 25mm wide and positioned edge-to-edge. The sheet that is produced is covered each side with veneer, usually beech, and then glued together under very high pressure. The glues that are used in blockboard are usually water-based and because of this it is only suitable for indoor use. Like plywood, because of its make-up, blockboard has good resistance to warping. If blockboard is cut, the edges can be a problem as they will not always match the hardwood veneer of the upper and lower surfaces. It may be necessary therefore to finish off unsightly edges with veneer strips.

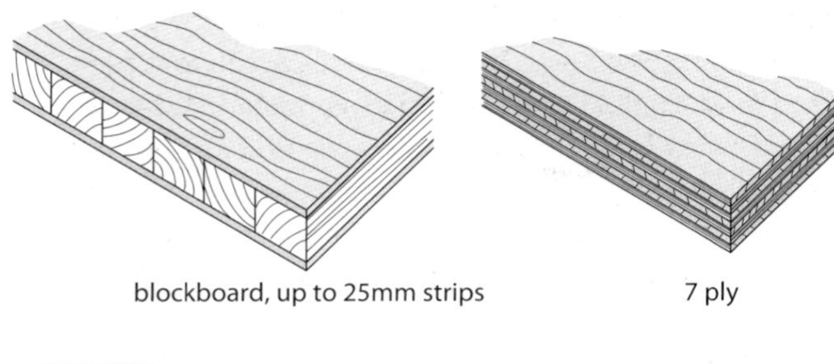

Figure 2.2 *Blockboard and 7 ply*

Modern materials and products

Over the last few years, modern materials and products have appeared in various situations and have been introduced into many products.

Thermo-ceramics

Thermo-ceramics are very advanced ceramic materials that have properties that make them particularly useful in some specialist engineering situations. They have an internal structure that makes them extremely hard and they are very stable at very high temperatures. These materials are used in places where there is a need for stability and strength at high temperatures, for example they are being used for the turbine blades in jet engines and they have been used in the turbo chargers of racing cars.

Thermo-ceramics are produced by combining ceramic and metallic powders by sintering. The powders are heated and then placed in a die and subjected to very high pressure until the particles bond with each other.

There are some disadvantages to thermo-ceramics. They can be brittle and can break if dropped and if there are imperfections in the ceramic material it can crack or break. This can be a real issue if the thermo-ceramic component is part of a complex system such as an aero engine. Added to these issues, the cost of thermo-ceramics can be significantly higher than more traditional materials.

Tinted glass and photochromic glass

Photochromic glass is glass that automatically darkens when exposed to ultraviolet light and then reverts to clear glass when the light source is reduced or removed. It is most commonly seen in spectacles that change from clear to dark in bright sunlight and then become clear when the sun goes in. The glass itself is impregnated with minute particles of silver halide. This is the same chemical that is in light-sensitive photographic paper. In photographic paper the silver halide reacts to light and an image appears on the paper. This is then fixed with chemicals. The fixing makes the process irreversible. In photo-chromic glass, because the silver halide is encapsulated and sealed within the glass, the process becomes reversible.

Solar panels

The generation of electricity using renewable resources is a topical subject and much effort is being put into finding alternative energy supplies. One method of generating electricity is to harness the light from the sun using photovoltaic, or solar cells. These cells have been long used in space exploration, but until recently have been regarded as too uneconomical to be used on a large scale on Earth. However, with the growing energy crisis, they are now being developed as one suitable alternative to the production of electricity using fossil fuels.

Photovoltaic cells are constructed from thin layers of silicon that have had various impurities added. When the cells are exposed to sunlight, one layer of the silicon becomes electron rich whilst the other becomes electron deficient. As a result of this electron flow, a voltage is set up between the two layers. If contacts are attached to the layers this voltage can be tapped. Unfortunately, the voltage produced is very small and to achieve a useful amount of power, large numbers of cells need to be connected.

At the present time the applications of photovoltaic power generation are somewhat limited. Solar cells are used in areas where mains power is not readily available such as marine buoys, on remote railway signalling or healthcare applications in developing countries. They are beginning to appear in domestic housing, but the cost of the installation can be fairly high.

WEBLINKS:

http://www.science.howstuffworks.com/solar-cell1.htm

This site offers a good insight into the construction and theory of the photovoltaic cell.

Liquid crystal displays

Liquid crystals are organic, carbon-based compounds that can exhibit both liquid and solid crystal characteristics. When a cell containing a liquid crystal has a voltage applied, and on which light falls, it appears to go 'dark'. This is caused by the molecular rearrangement within the liquid crystal. In the case of a digital clock or wristwatch, a liquid crystal display (LCD) has a pattern of conducting electrodes that is capable of displaying numbers via a seven-segment display. The numbers are made to appear on the LCD by applying a voltage to certain segments, which go dark in relation to the silvered background. As very small amounts of current are needed to power an LCD display they are ideal for portable electronic devices such as mobile phones, as battery life can be extended.

With the rapid advance of LCD technology came the full colour LCD display commonly used in laptops. In this case, each pixel is divided up into three sub-pixels with red, green or blue filters. By controlling and varying the voltage applied, the intensity of each sub-pixel can range over 256 colours.

Figure 2.3 *Liquid crystal display monitor*

LCDs have now evolved considerably and are at the forefront of modern domestic appliance technology with ever flatter, high-resolution LCD televisions and computer screens. However, there are some disadvantages to LCDs. The images produced on LCD televisions are, in some instances, inferior to those produced on traditional cathode-ray tubes and do not have the same viewing angle as traditional televisions.

Electroluminescent lighting

Electroluminescent (EL) lighting converts electrical energy into light (or luminescence) by applying a voltage across electrodes. An organic phosphor is sandwiched between two conductors and, as the electric current is applied, it rapidly charges phosphor crystals, which emit radiation in the form of visible light. EL lighting has extremely low power consumption, which makes it ideal for use as a backlight for LCD displays, for example the blue glow on a digital wristwatch. It is used to great effect in advertising to produce displays and posters with high visual impact.

Electroluminescent lighting is produced in the form of paper-thin wires, strips or panels, which are applied to designs that can then be animated by lighting up different areas of the poster. As they are so thin it is possible to utilise all the traditional methods of advertising, such as bus shelters, sides of buses and billboards as they are waterproof, highly visible and extremely reliable. The biggest disadvantage of EL lighting is that the organic phosphor materials used have a limited life span.

New and smart materials

Engineers are constantly developing new materials to meet specific requirements. Some of these new materials are becoming available to use in school. Smart materials are one of these new materials. A smart material is defined as a material that reacts to an outside stimulus, and then once that stimulus has been removed, changes back into its original state. The stimulus for changing a smart material is usually in the form of heat or electricity. Smart materials are now beginning to be used in a number of household and familiar products such as food processors, electric drills and in the control systems of central heating systems.

Shape memory alloys

A shape memory alloy is a material that can be deformed having been given an outside stimulus and then once that stimulus has been removed will revert back to its original state. The stimulus that makes some smart materials react is heat and this is usually, although not always, provided by an electrical input. Some others react to light, such as photochromic pigments and some to pressure, such as quartz. The most common and most readily available smart material is nitinol, an alloy of nickel and titanium. Nitinol is programmed to maintain a particular length or shape at a set temperature. If the temperature is raised the nitinol will contract or bend and it will remain in that state until the heat source is removed. As soon as the heat has been removed the material will return to its original state.

The remarkable thing about the material is that this cycle can be repeated millions of times. It has many applications including hot water systems, central heating systems, air conditioning units, fire alarms and seals for hydraulic systems.

However, there are some disadvantages to shape memory alloys. They are still relatively expensive to produce and are more difficult to machine when compared with more traditional materials such as steel. They also are not as strong as traditional materials when placed under similar loading – a piece of steel of the same dimensions will be able to contend with twisting or bending more than a shape memory alloy.

Reactive glass

Reactive glass is a smart glass. Unlike tinted glass which changes when sunlight falls on the impregnated silver-halide particles, reactive glass requires an external stimulus to make it turn from clear to dark. One application of reactive glass is in masks for electric arc welding. These have clear glass that remains clear until an arc is struck. When the light from the arc is sensed, the glass turns dark instantly. Unlike tinted glass, these welding shields contain electronics and batteries to enable them to operate. Another application is glass panels that can replace curtains or blinds and give privacy at the touch of a button.

Photochromic paint

Photochromic paints contain pigments that change colour according to light conditions. In one situation they might be one colour, but when sunlight or even ultraviolet light falls onto the pigment, there is an immediate and sometimes dramatic change in the colour of the pigment. These photochromic pigments can be mixed with base paint and can be used in many situations such as security markers or ultraviolet light warning sensors. Photochromic paints are reversible. When the light source or ultraviolet light is removed, the paint reverts to its original state.

Quantum tunnelling composites

Quantum tunnelling composite (QTC) is a new smart material that has many applications. It is a material that has some very interesting properties. When QTC is in its relaxed state, it is a near perfect electrical insulator but when it is stretched, compressed or twisted it becomes an electrical conductor and allows, in some instances, very high currents to pass through it. The greater the stress that is applied to the material the more it will conduct electricity. It is a material that has a vast number of practical applications. It is already being used in such products as power tool switches and robots. It also has potential applications in the textiles industry linking clothes to electronic devices.

WEBLINKS:

www.design-technology.info/alevelsubsite/page11.htm

A website that contains many links to interesting sites covering all aspects of smart materials.

THINK ABOUT THIS!

New and smart materials are becoming more commonplace. Where do you think new and smart materials could be used? Think about some everyday products that you are familiar with and see if any of the components could be replaced with smart materials.

Components

Nuts and bolts

Nuts and bolts are usually manufactured from low or medium carbon steel. The advantage of using nuts and bolts is that they can be easily undone, allowing components to be replaced or repaired.

The diameter of a bolt is always in millimetres and is sold in stock or standard sizes. The most commonly used thread form is the ISO metric thread. The metric thread is identified by the fact that the root angle of the thread is 60°. (Other thread forms have different root angles, for example, British Standard Whitworth threads have a root angle of 55°.)

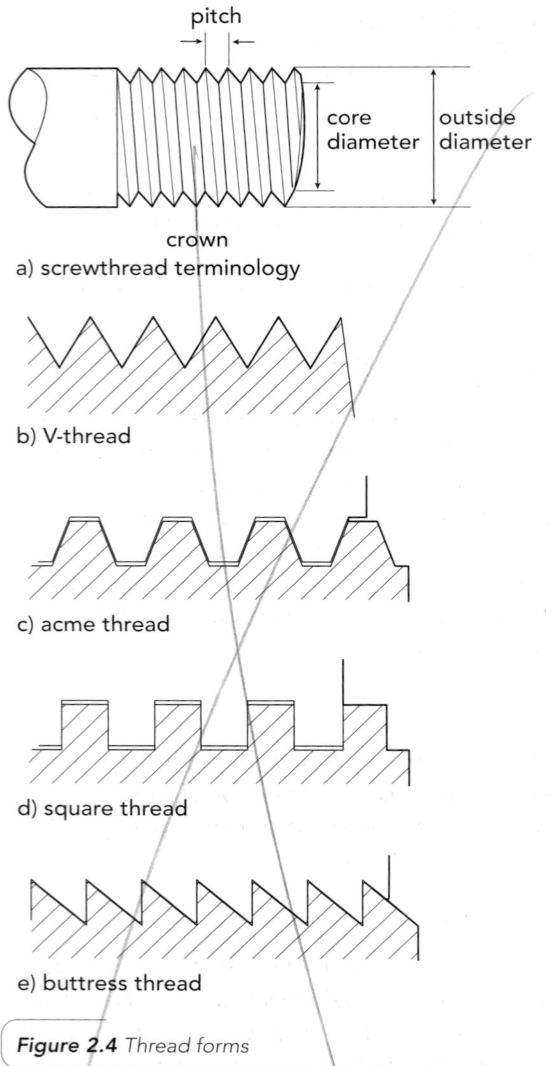

Figure 2.4 Thread forms

The other measurement that is often quoted when discussing threads is the pitch. The pitch of a thread is the distance that a screw or bolt will travel through one complete revolution. Although the most familiar thread form in nuts and bolts is the V form thread, there are others including square threads, buttress threads and acme threads.

Specialist nuts are used in particular situations. For example, where nuts and bolts are used in places where vibration is an issue, there is a danger that the vibration could cause the nuts to come loose. In these cases lock nuts could be used. These might be nuts with split pins positioned through them or nuts with nylon inserts which lock as the nut is tightened.

It is often necessary to cut a thread either into a hole, to enable a machine screw to be screwed into it, or onto a rod so that a nut can be screwed onto it.

Cutting threads in holes

The correct term for putting a thread into a hole is 'tapping'. The first consideration when tapping a hole is that the hole into which the thread is to be cut must be the correct size. The tapping hole diameter is determined by subtracting the depth of the thread from the outside diameter of the bolt or machine screw. For example, to produce a thread in a hole a 5mm bolt could screw into, a drill with a diameter of 4.2mm would be required. This would mean that the depth of thread would be 0.8mm. The tools required to cut the thread are 'taps' and a tap wrench.

There are three types of tap, the taper tap, the intermediate tap and the plug or bottoming tap, and each tap is used at different stages of cutting the thread. Once the hole is drilled the first tap, the taper, is placed into the hole and using a tap wrench the thread is carefully cut. It is important that the thread is cut carefully. The wrench and tap must be slowly turned through half a turn clockwise and then a quarter of a turn anticlockwise to break off the swarf. The tap will gradually work through the piece of metal until it appears on the other side.

During this process some lubrication should be applied. This lubrication traditionally has been tallow, but any grease or oil will work. This process is repeated with the intermediate tap and then finally with the plug or bottoming tap. Each tap takes out a little more material from the hole until the thread is completed.

One of the most important things to remember when tapping a hole is that the three taps must be at a right angle to the metal. If this is not done, when the thread is completed, any screw put into the hole will go in at an angle and appear wobbly when it is screwed in. Taps can be fragile and snap very easily. If a tap breaks in a piece of work it is often very difficult to remove it from the hole. The most difficult holes to put a thread in are 'blind holes', holes that do not go right through the metal. If you are not careful it is easy to keep turning the tap wrench into the blind hole and snap the tap.

Diameter of bolt/screw	Tapping drill diameter required
3mm	2.5mm
4mm	3.3mm
5mm	4.2mm
6mm	5.0mm
8mm	6.8mm
10mm	8.5mm
12mm	10.2mm

Table 2.10 Tapping drill sizes

Cutting threads on a rod

A thread can be cut on a rod either by hand or using a metalworking lathe. The most common method and by far the quickest way to put a thread on a bar is to cut it by hand. The tools required for cutting an external thread are a die and a die stock holder. Dies come in various stock or standard sizes and it is important that the size of the die matches the diameter of the bar. Therefore to cut a thread on a 10mm bar a 10mm die is required.

Dies are manufactured from High Speed Steel, which is harder than mild steel so will cut a thread on mild steel quite easily. Firstly, the end of the bar that is to be threaded needs to be slightly chamfered with a file. This will help the die to go onto the bar. The die then needs to be screwed onto the bar. The biggest problem is making sure that the die goes onto the work squarely. If this is not done a 'drunken thread' is produced. This means that any nut screwed onto the bar will not be at right angles to the bar and will appear wobbly on the bar.

One method of ensuring that the die is on the bar squarely is to use the tailstock of a centre lathe. Set the work up in the chuck on the lathe and place the die between the work and the tailstock and, *by hand*, turn the work until the die just begins to cut.

As the axes of the lathe and the tailstock are at 90° to each other, the die must go on at the correct angle. Once the die has started to cut the diestock should be gently turned half a turn clockwise and then a quarter of a turn anticlockwise to break off the swarf. As with cutting an external thread, some lubrication is again necessary.

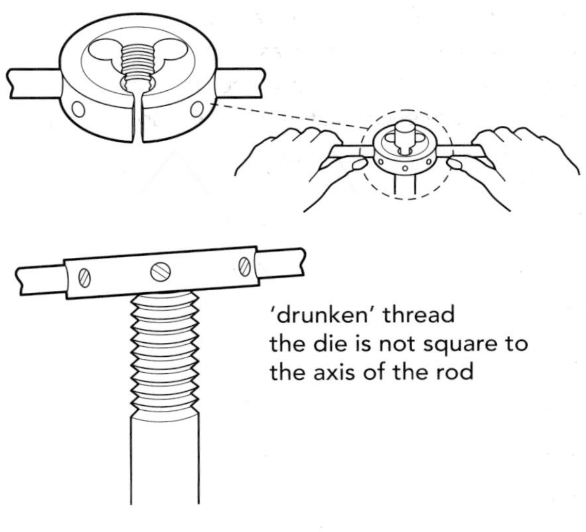
'drunken' thread
the die is not square to
the axis of the rod

Figure 2.5 *Thread cutting*

WEBLINKS:

www.nmri.go.jp/eng/khirata/metalwork/basic/bolt/index_e.html

This site gives in-depth information regarding the theory of threads and details of cutting threads in holes and on bars.

Spacers and washers

Sometimes in machinery, spacers and washers are used. Spacers are used in situations where components need to be separated on a shaft, such as on bicycle gears where it is sometimes necessary to put a spacer to separate the gears. Spacers come in a variety of materials including high density polyethylene (HDPE), nylon and steel. Washers can also be used as spacers, anti-vibration devices and locking devices.

Screws

Screws are often one part of a temporary joint. There are different types of screws that can be used in resistant materials. Machine screws are used in engineering and woodscrews are used for wood.

Machine screws

Machine screws have parallel sides with standard threads cut into them. They are usually manufactured from carbon steel and are mass produced. They can come with a variety of heads or tops. The most common are countersunk heads and cheese heads.

Countersunk screws are designed to fit into countersunk holes that enable the machine screw to be flush with the surface of the metal into which it is screwed. A cheese head is a head that is cylindrical in shape that sits above the metal into which it has been screwed. Added to those there are also specialist screws such as grub screws that might be used in particular engineering situations.

Screws can be tightened in a number of ways. The most common is a simple slot into which a screwdriver is placed. Other common types of screws can be tightened using Phillips and pozi-drive screwdrivers. In machines such as milling machines, drilling machines and lathes machine screws can be found that can be tightened by hexagonal Allen keys.

Wood screws

Wood screws are classified firstly by the shape of the screw head and secondly by the length of the screw itself. They come with three different types of head: countersink head, raised head and round head. Like machine countersunk screws, countersunk wood screws are designed to allow the screw to be flush with the surface of the wood. A raised head is similar to a countersunk screw except that the head itself is slightly domed.

A round headed screw is designed to rest on top of the surface of the wood. When screwing into wood it is advisable to make a pilot hole into which the screw is placed and screwed. If this is not done there is a danger of the wood splitting. A screw is driven in using a screwdriver. The shape of the screwdriver blade used depends on the shape of the screw head. It may be slotted, a Phillips or a pozi-drive head. Unlike machine screws, woodscrews are tapered. Woodscrews are usually sold by gauge and length.

Gauge	Metric
2	2.0mm
3	2.5mm
4	3.0mm
6	3.5mm
8	4.0mm
10	5.0mm
12	5.5mm
14	6.5mm

Table 2.11 *Screw gauge metric thread equivalents*

Figure 2.6 *Machine screws and wood screws*

Rivets

Riveting is a method of making a permanent joint in metal. There are two kinds of rivets, solid rivets and pop rivets. Solid rivets are manufactured from soft iron because they need to be ductile and easy to work as they have to be hammered into shape. Pop rivets are usually manufactured from aluminium and, instead of being hammered into shape, are fixed into place using a pop rivet gun.

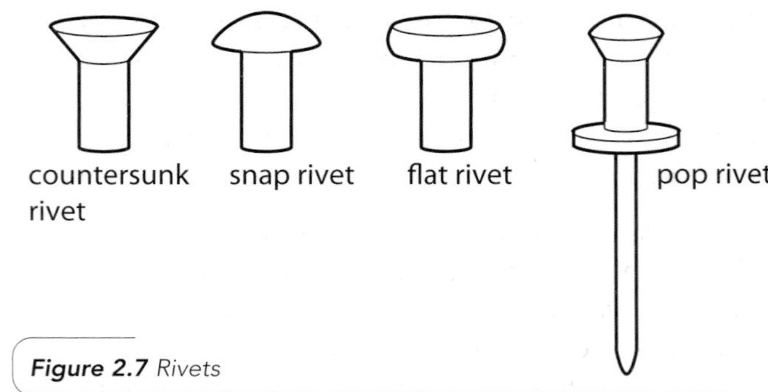

Figure 2.7 *Rivets*

Section 2.2 Industrial and Commercial Practice:
Joining techniques.

Gears

The theory of gears and gearboxes can be complex. Gear wheels are just one part of a mechanical system that when linked together can have a number of applications. Essentially, they are used to transmit rotary motion in various ways. Gears can be used to:

- change direction of rotary motion
- change spindle speeds
- transmit motion through 90°.

A gear is a wheel onto which a number of teeth have been cut. The teeth on a gear must be carefully designed. Gear teeth do not have flat sides because if they did the teeth on two gears would not mesh properly and probably jam. The shape of a gear tooth is in fact a mathematical curve called an involute. This type of shape enables the teeth on gears to mesh without the danger of jamming. It also enables forces to be transmitted between gears in the most efficient way.

Gears are linked together in gear trains. The simplest form of gear train consists of just two gears whose teeth mesh together. In this simple type of system, the input and the output are in different directions. If the gears are the same size, with the same number of teeth, the input and output revolutions would also be the same. If one gear was larger than the other, the rotational speed would be changed between each gear. If both gears were required to turn in the same direction, an extra gear, called an idler would be placed between the input and output gears. An idler can be of any size.

For more complex tasks, a compound gear train is used. This involves separate gear trains that mesh together. This could involve perhaps a number of gears positioned on the same shaft. Compound gear trains can significantly increase or decrease the number of possible rotational speeds.

As well as involving the change of speed and rotational direction, gear systems can also be used to change rotary motion into linear motion. Rack and pinion gear systems are found on the steering mechanisms of cars. A rack is a long piece of steel onto which gear profiles have been cut. A gear wheel is designed to mesh with the rack and as the gear wheel rotates it moves the rack from side to side. Another type of gear system that can be particularly useful in project work is the worm gear. A worm gear transmits forces through 90°.

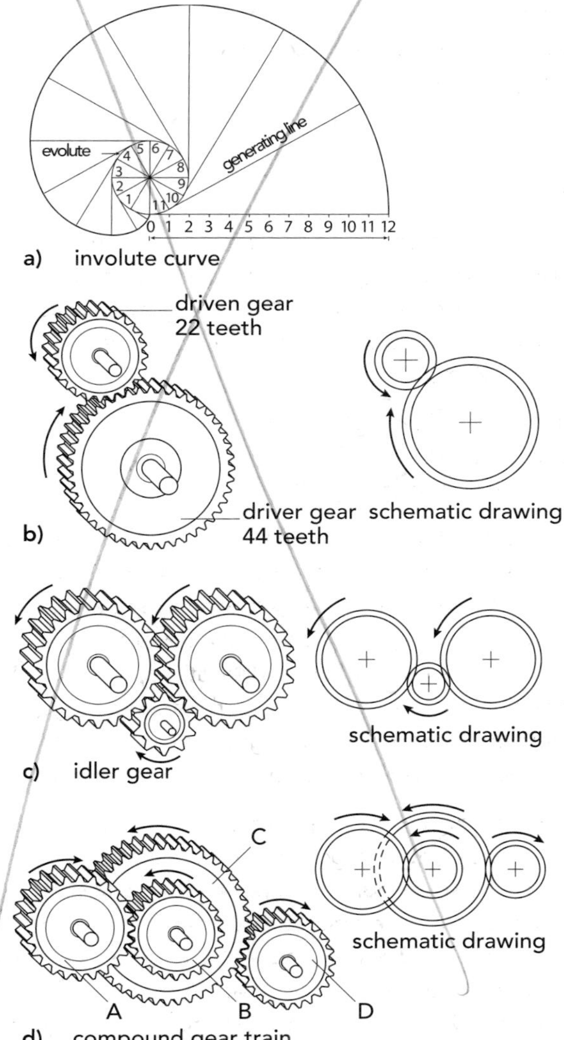

Figure 2.8 Gear trains and configurations

Spur gear	• Simple gearboxes • Speeds can be increased or decreased • Direction of rotation can be changed
Bevel and mitre	• Can have shafts at 90° • Different sizes of gears can change speeds • Commonly seen on hand drills
Worm and wheel	• Transmit force and motion through 90° • Reduction in rotational speed possible • Used in food mixers
Rack and pinion	• Used to change rotary motion into linear motion • Used in pillar drills • Used in steering mechanisms in cars

Table 2.12 *Application of common types of gear*

Bearings and bushes

A rotating shaft needs to be supported. However, where a rotating shaft is being supported, friction and wear will take place. In order to reduce the effects and level of friction, some kind of bearing is usually required. Bearings reduce the effects of friction and allow shafts to spin more efficiently.

Plain bearings

A plain bearing in its simplest form can be a support that holds a rotating shaft. However, because of friction the material from which the support is manufactured would soon wear down and either cause the shaft to spin inefficiently or cause a catastrophic failure of the support. One way of solving this problem is to place between the support and the shaft a sleeve of some kind of 'bearing material' that has a low coefficient of friction. A sleeve of bearing material is called a bush.

The most common material for bushes is bronze, which is an alloy made up of approximately 90 per cent copper and 10 per cent tin. The disadvantage of using a bush is that they wear out over time and have to be removed and replaced. Bushes can be manufactured from other materials. For instance, nylon is increasingly used as it is a 'self-lubricating' material and some ceramics are being used. Ceramic bearings are produced using a sintering process and the advantage of using sintered ceramics in bearings is that they are porous and if saturated with oil will soak up the oil and thus provide lubrication for the bearing.

Journal bearings

A journal bearing is in some ways similar to a plain bearing. Its main difference is that, instead of having a bush insert, it has a layer of grease, or oil, between the rotating shaft and the support. The theory is that the layer of grease, or oil, is very thick and therefore the shaft never actually comes in contact with the support. In order to keep this type of bearing working efficiently, it is important that regular servicing is undertaken and the grease is kept to the correct level. To top up the oil or grease, it is fed into the bearing under pressure through a small hole.

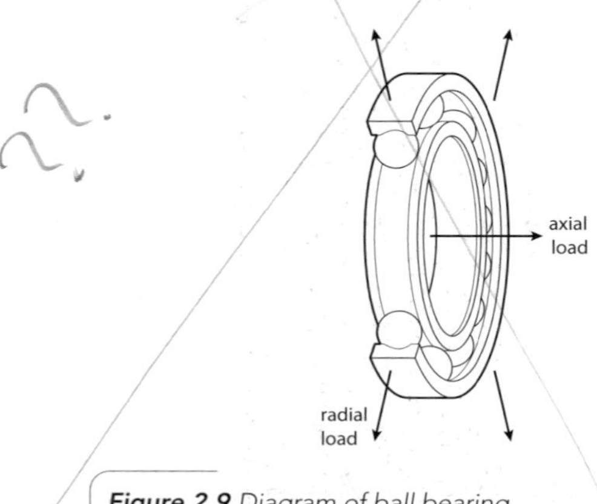

Figure 2.9 *Diagram of ball bearing*

Ball bearings

Ball bearings are one of the commonest forms of bearing. Essentially, spherical steel balls are held in a ball race that allows the inner race to spin independently from the outer race.

Cams and followers

Cams are usually used in machines where, in some part of that machine, rotary motion needs to be changed into reciprocating linear motion. Cams come in a wide variety of shapes, all of which are designed for specific tasks. Above the cam and resting on it is the cam follower. As the cam rotates, the follower is pushed up and then, following the profile of the cam, falls back to its original position.

The simplest form of cam is circular and is often referred to as an eccentric cam. It is essentially a circular disc that is fitted to a driving shaft off centre. This type of cam gives the follower a very smooth rise and fall. In some instances more complicated movements are required such as where cams are used in internal combustion engines to control the opening and closing of the inlet and exhaust valves. It may be that the valves need to be held open for a set amount of time. If this is the case then the cam needs to be designed to allow this to happen. A cam might therefore have a curve which pushes the follower up, then have a circular section which allows the follower to dwell at the highest point and then have a curve that allows the follower to drop down to its lowest point. There are three types of movement that can be created by cams:

- uniform velocity
- simple harmonic motion
- uniform acceleration and retardation.

Each type of these movements requires a different cam profile. Depending on the situation where cams are required there are also a number of cam follower types that can be used. The most common types of follower are flat-foot, knife edge and roller.

Type	Profile	Description
Flat		Lots of friction Cannot follow hollow contours
Knife		Provides the most accurate conversion of movement Can be used to follow hollow contours
Roller		Least friction Cannot be used to follow hollow contours Most expensive

Figure 2.10 *Diagram of followers*

Type of cam	Characteristics
Pear-shaped cam	• Used to control inlet and outlet valves in engines • Symmetrically shaped so rise and fall times are identical
Eccentric or circular cams	• Simplest type of cam as based on an off-centre circular disc • Smooth rise and fall • Constant acceleration and retardation • Used in fuel pumps
Heart-shaped cams	• Symmetrical cams that generate continuous motion • Used in sewing machines and to wind relay coils and motor windings
Snail cams	• Generate output with uniform acceleration and retardation

Table 2.13 *Cam characteristics*

Industrial and commercial practice

Getting started

As a designer, you need to know how products are manufactured. It's quite simple – you need to be able to design products that can be made and to make products that you have designed. What would happen if designers simply scribbled something on a piece of paper and gave it to somebody else to make? Would the manufacturer understand the drawing and, more importantly, what might go wrong if information is not communicated effectively?

Scale of production

The scale of production is an important factor to be considered when developing any product. It has an impact upon all design and manufacturing decisions, including:

- the number of products or units manufactured
- the choice of materials and components
- the manufacturing processes, speed of production and availability of machinery and labour
- production planning, the use of just-in-time (JIT) and stock control, including the use of information and communication technology (ICT) systems
- production costs, including the benefits of bulk buying, the use of standard components and eventual retail price.

One-off

One-off production is often referred to as job production and relates to 'tailor-made', bespoke or customised solutions. A key feature of one-off production is a single, often high-cost product that is manufactured to a client's specification. This kind of one-off product may be relatively high-cost because a premium has to be paid for any unique features, more expensive or exclusive materials and time-consuming handcrafted production and finishing.

Batch

Batch production involves the manufacture of identical products in specified, predetermined 'batches', which can vary from tens to thousands. A key feature of batch production is flexibility of tooling, machinery and workforce to enable fast turnaround, so production can be quickly adapted to manufacture a different product, depending on demand. Batch production often makes use of flexible manufacturing systems (FMS) to enable companies to be competitive and efficient.

Figure 2.11 *Hand-made bespoke chair*

The use of computer-integrated manufacturing (CIM) systems involving automated machinery enables production 'downtime' to be kept to a minimum.

Batch production results in lower unit cost than one-off production. Economies of scale in buying materials enable cost savings and identical batches of consistently high-quality products are manufactured at a competitive cost.

Mass

Mass production (or high-volume production) of most consumer products makes use of efficient automated manufacturing processes and a largely unskilled workforce. Mass-produced products are designed to follow mass market trends, so the product appeals to a wide national and international target market. Production planning and quality control (QC) in production enables the manufacture of identical products. Production costs are kept as low as possible so the product will provide value for money.

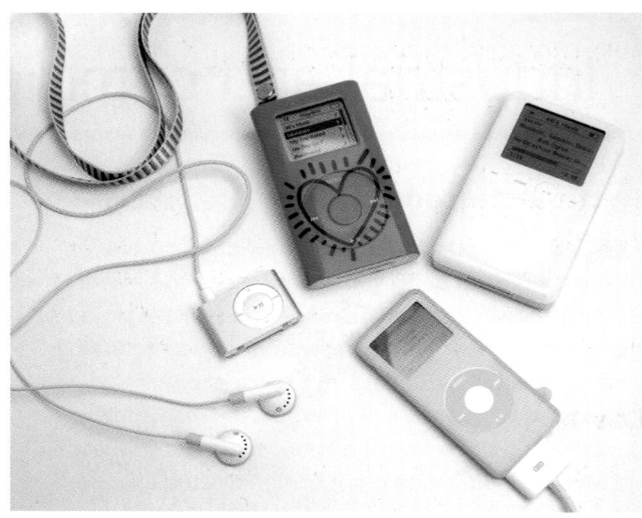

Figure 2.12 *Mass-produced products*

Continuous

Continuous production is used to manufacture standardised mass-produced products that meet everyday mass-market demand. The production of a blow-moulded fizzy drinks bottle, for example, necessitates 24-hour production, 7 days a week to satisfy consumer demands for soft drinks. This type of production is highly automated and uses machines that can run continuously for long periods of time with breaks only for routine maintenance.

Scale of production	Advantages	Disadvantages	Applications
One-off	• Made to exact personal specifications • Highly skilled craftsperson ensures high-quality product	• Expensive final product in comparison to larger scales of production • Generally labour intensive and can be a relatively time-consuming process	Bespoke piece of furniture for particular situation
Batch	• Flexibility in adapting production to another product • Fast response to market trends • Very good economies of scale in bulk buying of materials • Lower unit costs than one-off produced products	• Poor production planning can result in large quantities of products having to be stored, incurring storage costs • Frequent changes in production can cause costly re-tooling, reflected in retail price	Seasonal garden furniture
Mass	• Highly automated and efficient manufacturing processes • Rigorous quality control ensures identical goods • Excellent economies of scale in bulk buying of materials • Increased production means that set-up costs are quickly recovered • Low unit costs • Reduced labour costs	• High initial set-up costs due to very expensive machinery and tooling needs • Inflexible; cannot respond quickly to market trends	Electronic products, e.g. mobile phones and games consoles, commercial packaging
Continuous	• As mass production • Extremely low unit costs • Runs continuously 24 hours, 7 days a week	• As mass production • Very little flexibility at all as production set up 24/7	Cans and bottles for the drinks industry

Table 2.14 *Scale of production*

Material processing and forming techniques

It is important that you have a sound understanding of the material processes and forming techniques that are an essential part of Resistant Materials Technology. Your exams will contain questions about these processes and techniques and they may well be useful to you when you are working on the coursework element of the course.

Casting

Casting is the pouring, or forcing, of molten metal into a cavity and then allowing the metal to cool and solidify. The two principal types of casting metal are sand casting and die casting.

Sand casting

Sand casting is the process where molten metal is poured into a sand mould that contains a cavity of the desired shape of an object or component. Once the metal has solidified, the sand is broken open and the product or component is removed.

The first step in sand casting is to make a mould in sand using a pattern. A pattern, usually manufactured from wood or medium density fibreboard (MDF), is a replica of the object that is to be cast. It is identical to the finished product apart from the fact that it is slightly oversized to allow for the contraction of metal as it cools down. The majority of patterns are made in two parts. They are cut down the centre line and are held together using dowels.

Using split patterns allows two sides of a sand mould to be created and complex casting to be produced. Pattern making for casting requires great skill. The pattern must be designed with sloping sides so that it can be easily removed from the mould without damaging the walls of sand. It must have no undercuts or sharp edges and any corners must be finished off with 'fillet' radii. The final pattern is usually finished off in gloss paint. Some patterns are very complex and involve them being made in a number of pieces.

Different types of sand are used in casting, the most common being 'green sand' and 'Petrabond®. Green sand uses water to enable it to bond together and can be reused, but Petrabond® is oil based and can only be used once. Before using the sand it needs to be sieved and any lumps removed. Green sand needs to be slightly damp before it can be used. It should keep its shape within a clenched fist.

However, it should be remembered that, before pouring molten metal into a mould made from green sand the sand must be thoroughly dried out. If molten metal is poured into a damp or wet mould, there is a real danger that the mould will explode. The mould in sand casting is produced in two open boxes called the 'cope' and the 'drag'.

The production of the mould can be broken down into a number of stages:

Stage one

The drag, the bottom half of the mould, is placed upside down and one half of the pattern is placed face down in the centre of the drag. Parting powder or French chalk is sprinkled over the pattern. This is done to stop the sand sticking to the pattern when the pattern is removed. Sieved sand, or facing sand, is then packed around the pattern. Facing sand needs to be sieved, so that it can mould easily around the pattern. This facing sand should cover the pattern to a depth of about 30mm. The sand should be rammed down to compact it. Once the facing sand has been put around the pattern the drag is filled with unsieved or backing sand. The sand should be rammed down and then levelled off. If the process has been done correctly, the drag can be lifted and moved without the sand falling out.

Stage two

The drag is then turned over and the cope placed on the drag. The second half of the pattern is placed on the first half which is embedded in the sand in the drag. More parting powder is sprinkled over the pattern and the 'sprue pins' are placed into the sand. Sprue pins will make the cavity in the sand where the molten metal is poured in and where air is released from the cavity. The cope is then filled with facing sand and then backing sand. Once this has been done, hollows are put around the tops of the sprue pins. These hollows in the sand will help the molten metal to flow freely into the mould. The sprue pins are then carefully removed and the cope and drag separated.

Gates or channels are cut between the pattern and the holes created by the sprue pins *before* the pattern is removed. Once the gates and channels have been cut, the pattern is removed and the cope and drag is reassembled, placed on a bed of sand and made ready for casting. It is important to remember that, if green sand has been used to create the mould, then it must be left for two or three days to thoroughly dry out to avoid the risk of explosion.

Stage three

The metal to be poured into the mould is heated in a crucible in a furnace. When the metal becomes molten it is fluxed and degassed. This process removes any impurities and helps release gases that may have formed during the process. If this is not done, impurities can spoil the final casting and gases can cause bubbles to be formed in the metal. The molten metal is then poured slowly and evenly into the holes that were created by the sprue pins.

The hole that the metal is poured into is referred to as the 'runner' and the hole where the air is released is known as the 'riser'. When the molten metal appears in the riser the pouring should stop. The metal is allowed to cool and solidify. Once cooled, the cope and drag can be separated and the sand broken open and the component removed. When the component has been removed it will need to be 'fettled'. This is where the runners, risers and flashings or waste material are removed using a hacksaw.

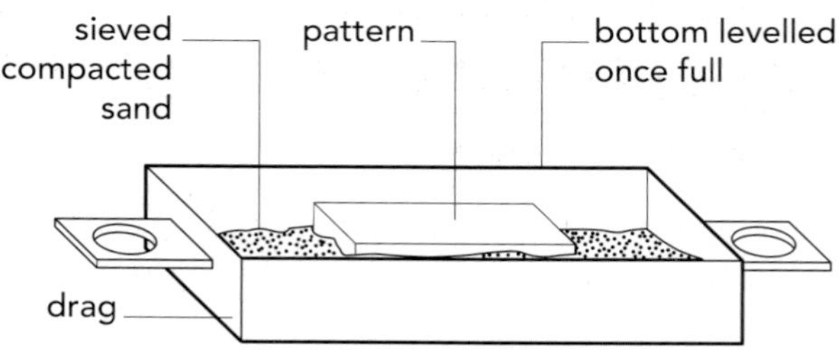

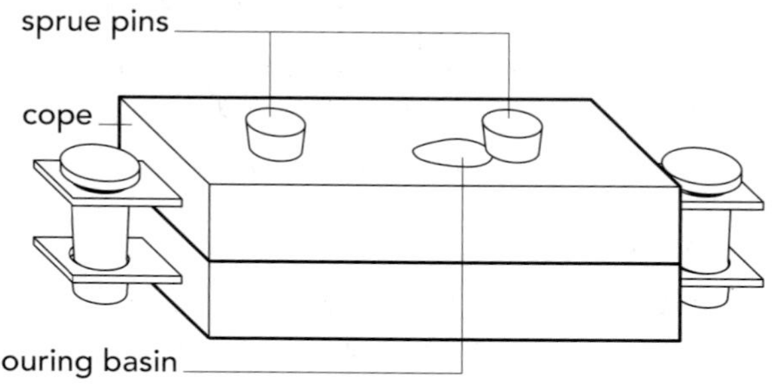

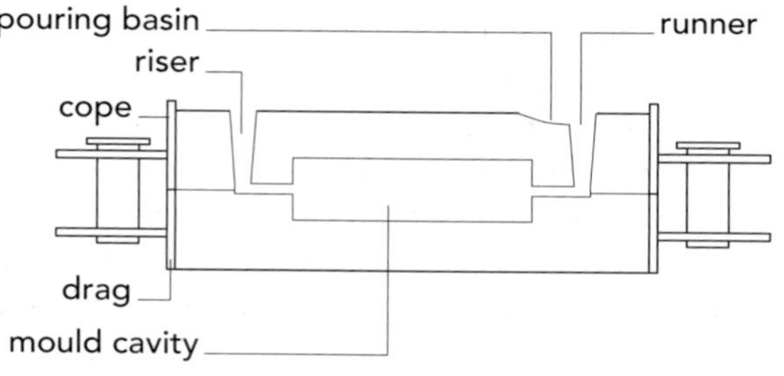

Figure 2.13 *Diagram of sand casting*

Die casting

This is a process for producing metal products, usually in non-ferrous metal, by either forcing molten metal under high pressure into a die, a process known as pressure die casting, or allowing molten metal to flow into a mould by gravity, which is known as gravity die casting. Dies are reusable moulds that are manufactured from steel. Die casting is a high-volume, mass-production process and is used when a large number of products are required.

In pressure die casting, the metal is forced into the alloy steel dies under hydraulic pressure. The process is broadly similar to injection moulding in plastics. The process is highly automated and produces high-quality castings. Unlike sand casting, this process cannot be undertaken in the school or college workshop.

There are a number of steps in pressure die casting:

- Mould is sprayed with lubricant. This helps to control the temperature of the die and helps with the removal of the finished casting.
- Molten metal is then shot under high pressure into the die.
- When the die is filled, the pressure is maintained until the metal has solidified.
- The die is opened and the casting is removed by ejector pins.
- Finally the gates, runners and risers are removed.

Unlike sand casting, die casting can produce multiple castings, producing a number of castings with a single cycle of operation.

In gravity die casting, the molten metal is poured into the mould and, as the name suggests, gravity helps the metal to flow around the mould.

FACTFILE:

Name	Advantages	Disadvantages	Applications
Sand casting	• Inexpensive • Complex shapes can be produced • Large components can be produced	• Sand moulds can only be used once • Surface finish not always good • Labour intensive • Slow production rate	• Engine blocks • Garden furniture • Caterpillar tracks
Die casting	• High production rate • Good surface finish • Economical for large production runs • Good control over mould temperature during process • Elements such as screw threads can be included in the casting	• High set-up costs • Long lead time • Limited sizes possible • Not all alloys suitable for the process • Large volume production required to make process economical	• Taps • Model cars

Milling and routing

Milling is the process of cutting away metal, by feeding a piece of work past a rotating cutter. Routing is a similar process but usually associated with wood, composites and plastics. The most common operation that milling and routing is used for is cutting slots into material.

There are two types of milling machine, vertical and horizontal. The vertical milling machine looks similar to the drilling machine, but instead of using a drill, it has a rotating cutter that can be raised and lowered. Work is clamped to the table below the cutter. The table is able to move in two directions and the cutter can be moved up and down. This allows slots to be cut either along or across a piece of metal to the desired depth. With the horizontal milling machine the work is similarly clamped onto a table that can move in two directions but instead of the cutter turning with its axis being vertical, the axis of the mill is horizontal. When using a horizontal milling machine it is usual practice to 'upcut' mill. This means that the work must be moved in the opposite direction to the rotation of the cutter. The work must be securely fastened to the machine and lined up with the cutters to enable the slots to be milled in the correct position. Two important points to consider when milling are the rotational speed of the milling cutter and the speed at which the work passes the cutter.

When routing slots in wood, the work is usually secured to a bench and the router is hand-held. Attached to the router is a "fence". The fence helps to guide the router to ensure that slots are cut in the correct position parallel to the edge of the work. It is therefore very important that before routing, the edges of the wood have been carefully planed to ensure that they are square. As with all processes involving power tools and machinery, it is important that the correct safety measures are adhered to.

WEBLINKS:

www.technologystudent.com/equip1/hmill1.htm

www.technologystudent.com/equip1/vert1.htm

These pages have useful explanations and diagrams of vertical and horizontal milling machines.

Figure 2.14 *Vertical milling machine*

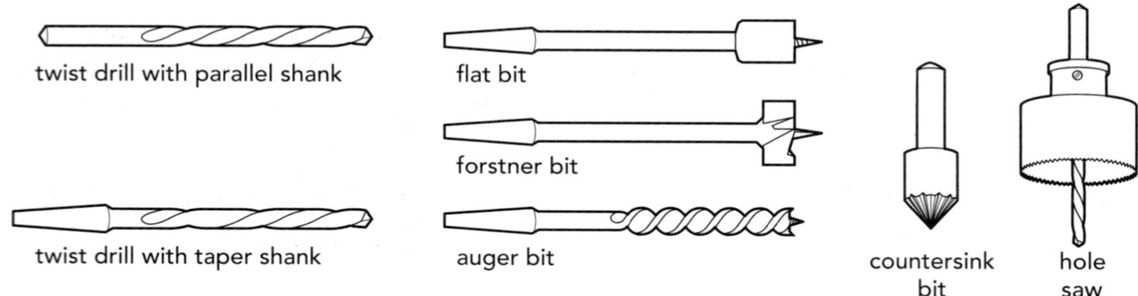

Figure 2.15 *Drill bits*

Drilling

Drilling is the process of making holes by using a rotating cutting tool that is secured in either a hand operated drill or a drilling machine. The most common type of drill is the twist drill. Twist drills are usually manufactured from high speed steel and can be used to produce holes in most materials.

There are two types of twist drills – parallel shank and taper shank drills. Parallel shank drills are held in the drilling machine by a chuck and taper shank drills are placed directly into the machine and held through friction. Drill bits have fluting or grooves along them. These flutes enable the swarf, or waste material, to be carried away from the tip of the drill bit whilst cutting through the material. When drilling metals, lubricants or cutting fluids should be used to keep the tip of the drill cool and keep the drilling efficient.

Other types of drills found in the workshop are:

- Flat bits – used to drill deep holes in wood.
- Forstner bits – used to drill flat-bottomed holes in wood.
- Auger bits – used to drill deep holes using a carpenter's brace.
- Countersink bits – conical drill bits, usually 90°, that when used allow a countersunk head screw to be inserted so that the head of the screw is flush with the surface of the wood or metal.
- Hole saws – saws that are circular shaped cutting rings that range from 20mm to 150mm in diameter.
- Tank cutters – circular cutters that have an adjustable radius for cutting holes in sheet material.

Because wood is generally softer than metals, drilling into wood tends to be easier and faster than drilling into metals. When drilling into wood, cutting fluids are not used. One issue that is often apparent when drilling wood is that wood can, because of the friction between the drill and the material, overheat and burn. It is therefore important to use a sharp drill set to the correct rotational speed. When marking out for drilling it is important to establish the correct position. The best way is to either use a scriber on metal or a sharp pencil on wood to draw lines which cross where the hole is to be drilled. When drilling into metal it is also necessary to centre punch where the hole is to be so as to guide the drill into the correct position.

Turning

Turning can be undertaken using either a metalworking centre lathe or a woodturning lathe. The two basic processes that the lathe can perform are facing off, the smoothing of the end of a piece of material and turning down, reducing the diameter of a piece of material. Unlike most other processes, where work is securely fixed or bolted down, the work on a lathe rotates whilst it is machined. On a metalworking centre lathe, the cutting tools are securely fixed to the lathe whereas, with a woodworking lathe, the cutting tools are held in the hand and are rested on a toolrest.

When turning on a metalworking centre lathe the work is usually securely held in a chuck. The chuck is rotated and a cutting tool moved along the work to reduce the diameter of the work. When wood turning, the work being machined is usually secured to a faceplate or turned between centres.

There are two types of chuck – the three jaw self-centring chuck, for holding round or hexagonal bars, and the four jaw independent chuck for holding material of any cross section. In metal turning, the cutting tool is secured in a tool post. The movement of the tool post is controlled by hand wheels. The cutting tool can be moved along the work (turning down) to reduce its diameter or across the work (facing off) to smooth down its end. With woodturning, the cutting tool is positioned on top of the tool rest and unlike metal turning the cutting tool is hand-held throughout the turning operation. In woodturning, the cutting tool is placed on the tool rest and then carefully moved along the work.

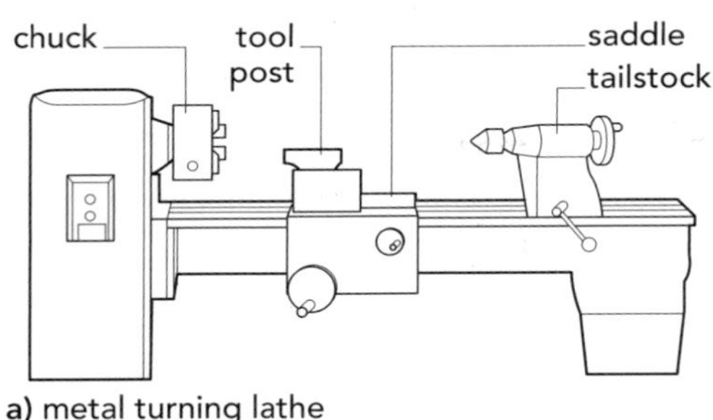

a) metal turning lathe

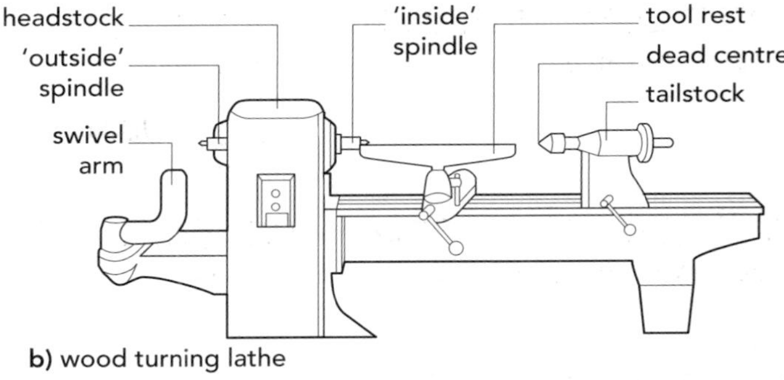

b) wood turning lathe

Figure 2.16 Turning

Care must taken when setting up the lathe. With metalworking, the work must be secure in the chuck and aligned correctly with the cutting tool positioned at centre height. If the tool is too high it will cut very inefficiently and damage the tool, or if it is too low the work will attempt to ride up over the tool and snap the cutting tip. When turning wood, care must be taken to ensure that the timber is free from knots.

Metalworking centre lathe

The metalworking centre lathe is probably the most important and versatile machine in the workshop. It can be used for a large number of tasks, from turning down diameters on metal bars to drilling holes in the centre of round material, producing tapers or producing threads on bars, among others.

Centre drilling and boring

Centre drilling is when the lathe is used to drill a hole in the end of a rod or bar along its axis. Unlike using a drilling machine where the work is secured and the drill rotates, centre drilling involves the work rotating in the chuck with the drill held stationary in the tailstock. This process must be carried out with care. Before drilling the required hole, a smaller hole must be produced using a centre drill. Once the centre drill has been used, it is changed for a twist drill and the hole completed. The hole produced by the centre drill guides the twist drill into the work. If a large hole is required it may be necessary to use a boring tool. This is similar to the cutting tool used to turn down the outside of a metal bar, but is used internally to enlarge holes. This is held in the tool post and is again controlled by hand wheels.

Whilst carrying out all these operations on the metalworking lathe, there is a danger of heat building up due to the friction of the cutting tool against the metal or, in the case of drilling, at the tip of the twist drill. If the heat is allowed to build up too much the cutting tool, or drill, bit can become blunt and inefficient. It is therefore important to keep the work cool. To do this a coolant, in the form of soluble oil, is applied to the work. In commercial lathes, coolant is usually pumped onto the work in order to keep it cool. However, in the school workshop coolant is often applied using a brush. It is important to remember that in order to keep the end of a drill cool, the drill needs to be removed from the hole being drilled. This also helps keep the tip of the drill free from swarf and again makes the cutting more efficient.

Tool	Application
Right-hand knife tool	• Used to face the left-hand edge or cut a shoulder to the left
Left-hand knife tool	• Used to face the right-hand edge or cut a shoulder to the right
Round nose tool	• Cut in any direction and produce radii
Parting off tool	• Moves at right angles into the work to sever material
Form tool	• Special profiles to cut specific shapes
Knurling tool	• Produce pattern or textured surface on the material
Boring tool	• Tool to enlarge existing hole when large drill unavailable or to produce a flat-bottomed hole

Table 2.15 Metalworking centre lathe tools

When using the lathe, it is important that the correct 'speeds' and 'feeds' are established. The speed refers to the speed of the spindle at which the work will turn in terms of revolutions per minute (RPM). The feed relates to the rate at which the cutting tool moves along the work as it cuts.

Another factor that must also be taken into account is the 'depth of cut'. The depth of cut refers to how deep the cutting tool goes into the work as it moves along the material. The deeper the depth of cut the more material will be removed but the friction will be greater, thus making the cutting more inefficient. There are formulas that enable the operator to establish the correct speeds and feeds when setting up a lathe, but in general terms, the larger the piece of work to be turned, the slower the speed and feed and the smaller the depth of cut should be.

Producing a screw thread using a centre lathe

In the school workshop, the cutting of threads is usually undertaken by hand using taps and dies. However, external threads can be cut using a centre lathe. A cutting tool ground to the correct thread form angle is set up in the tool post and is used to cut along the bar to produce the correct thread profile on the bar.

There are two factors that are vital. Firstly, the cutting tool profile must match the thread profile required. With a metric thread this means that the angle of the tool must be 60°. With other thread forms, this angle will vary. Secondly, the rate of rotation of the work must be in relation to the longitudinal cut in order to create correct thread pitch. It is important that with every cut, the tool comes in contact with the work at exactly the same point. In order to cut a thread correctly, a lathe with a 'lead screw' is used. This ensures, by using gears, that the rotation of the chuck is directly related to the speed at which the tool moves along the work.

Knurling

Knurling is when the lathe is used to put a pattern onto the surface of the material, for example it may be that on the end of a particular tool a textured grip is required. This is often in the shape of a diamond pattern and is produced using a knurling tool. The knurling tool consists of two hardened steel rollers that have the pattern to be cut engraved on them. The pattern can come in two forms, straight or diamond shaped. As the work in the lathe rotates, the rollers on the knurling tool are pushed into the work. As the steel rollers are harder than the bar being knurled a pattern is imprinted on the bar. This operation is usually undertaken on a slow speed.

Parting off

Parting off is usually the final stage of working on the centre lathe. This is where a special tool is put into the tool post and is used to cut right through a piece of work, thus severing it. It is vital that the parting off tool is kept sharp with correct angles ground on it. It should be set exactly at centre height and fed into the work carefully at a uniform rate. In industry, the speed of rotation during parting off can be quite high, but in the school workshop it is usually kept fairly low. When parting off, care should be taken to ensure that the component being cut does not fly off the machine.

Wood lathe

The woodworking lathe is, in many ways, similar to the metalworking centre lathe. Work is not held in a chuck but is secured either onto a faceplate for turning items such as bowls, or where long pieces of work such as spindles or chair legs need to be turned, between the centres of the head- and tailstocks. The turning process is very similar to working on the metalworking lathe except that when woodturning, tools are held in the hand against a rest and not in a tool post.

There are three basic woodturning tools:

- gouges
- chisels
- scrapers.

Before woodturning, it is important that the timber is thoroughly prepared. If turning down long pieces of timber, the corners should be planed down. If small pieces are being turned, such as bowls on a faceplate, the corners should be cut off. Any knots should be avoided as these can get caught by the turning tool and fly out. If a long item is being turned between centres, the centre of the material should be carefully established so that when it is secured to the lathe and rotated it is not unbalanced. The work is held on the lathe between a driven centre, which is positioned in the headstock and a rotating centre in the tailstock. Once the work has been secured onto the machine the tool should be checked for sharpness. The tool post should be positioned so that the top of the turning tool is at centre height of the work. As with the metal lathe, it is important to set the correct spindle speed for the task in hand.

THINK ABOUT THIS!

Safety is always a key issue when using any kind of workshop machinery. Look at one of the machines in your school or college workshop and complete a risk assessment for a particular procedure.

WEBLINKS:

www.minilathe.com

This is an interesting site that, although dealing with a specific lathe, has a great deal of useful information.

Injection moulding

Injection moulding can be used for both thermoplastics and thermosetting plastics. However, because thermosetting plastics are created using resin and a catalyst, thermoplastics are the most common material used in injection moulding.

The process involves molten plastic being injected under pressure into a mould. Plastic granules are placed in a hopper and then forced into a heated chamber using an Archimedean screw. The plastic is then injected into a two-part mould which is a 'negative' of the product required. The component is rapidly cooled (usually with water) and then ejected from the mould. The moulds are usually made from steel and are high-quality precision made using spark erosion. The task of designing and manufacturing moulds for injection moulding is highly skilled. As well as including all the detail of the product, the mould must be constructed in such a way so as to allow the molten plastic to be injected into all the recesses of the mould and it should be easy to eject from the machine.

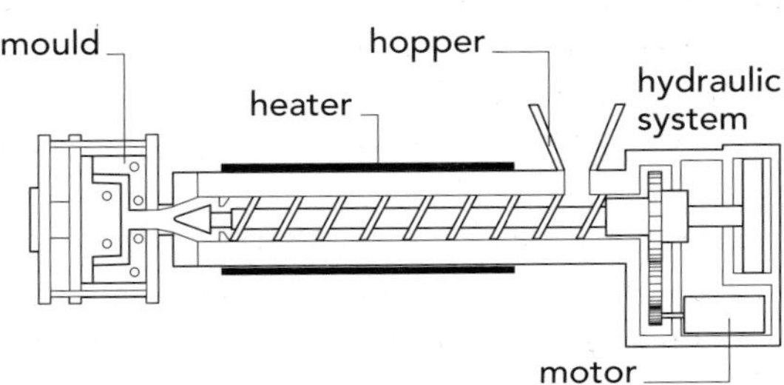

Figure 2.17 Injection moulding

Injection moulding is widely used in industry and is used for manufacturing a wide variety of products, from bottle tops to large items such as patio furniture. The commonest materials used in injection moulding are thermoplastics such as acrylonitrile butadiene styrene (ABS), nylon and polyethylene.

WEBLINKS:

www.design-technology.org/injectionmoulding2.htm

An animated explanation of injection moulding.

Extrusion

Extrusion is a manufacturing process that is used to create long products that have a particular cross section. The long component parts of unplasticised polyvinyl chloride (uPVC) window frames and rainwater gutters are examples. The process is very similar to injection moulding. The material is heated up in the same way and then, instead of being forced into a mould, is forced through a die of the desired cross section. Hollow sections can be produced by placing a pin inside the die. Extrusions may be continuous, which produces indefinitely long material such as guttering, or may be semi-continuous, producing many short pieces.

Blow moulding

Blow moulding is the process used to produce hollow products such as bottles from thermoplastics. The first part of the blow moulding process is similar to injection moulding and extrusion. The plastic, in the form of granules, is fed into a hopper. Like injection moulding, an Archimedean spiral feeds the granules through a heated chamber.

As the plastic is heated it begins to soften and is extruded into a hollow tube. This hollow tube is called a parison. The parison is then clamped in a metal mould and air pumped into it. The parison inflates to the shape of the mould. Once the plastic has cooled the mould is opened and the bottle is ejected.

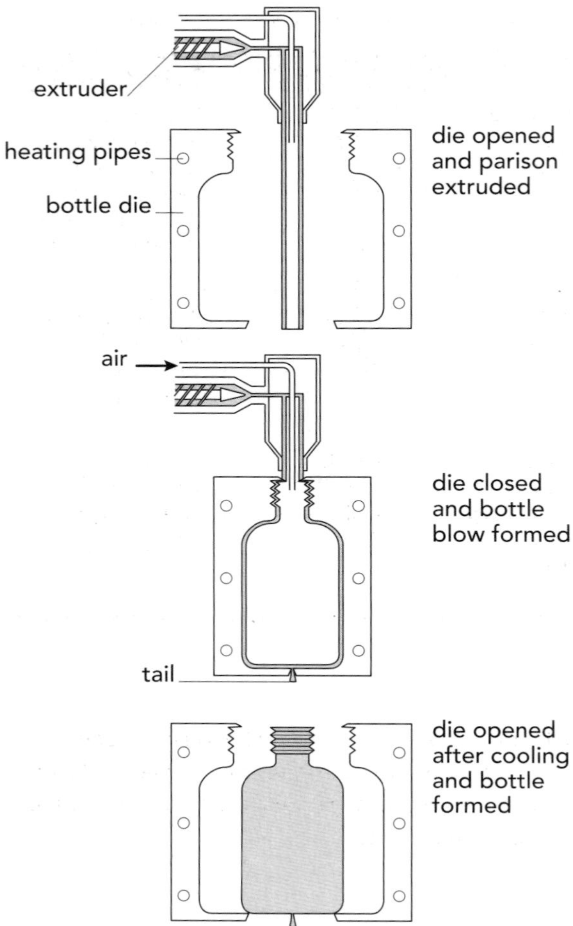

Figure 2.18 *Blow moulding*

WEBLINKS:

www.technologystudent.com/equip1/blowm1.htm

An animated demonstration of the blow moulding process.

Vacuum forming

The vacuum forming process involves forming thermoplastic sheets into desired shapes by firstly applying heat to the plastic to make it soft and pliable so a vacuum is produced and air pressure forces the sheet of plastic around the former. The former is a replica, made from wood, MDF or similar material, of the item that is to be produced. Once the thermoplastic has been formed it cools and is then removed. Vacuum forming has many applications, from packaging in the food industry to the acrylic bath in your bathroom.

Stages of vacuum forming

Stage 1

Initially a high-quality former needs to be produced. It should be the exact shape of the item to be vacuum formed. The sides of the former should include at least a one-degree taper to ensure that the final vacuum formed product can be easily removed and it should contain no undercut elements. There should also be a number of very small holes drilled in the former so that air can be sucked out during the vacuum forming process.

Stage 2

The former is placed onto the machine and a sheet of thermoplastic is clamped into place over the former. Heat is applied to the thermoplastic in order to make the plastic soft and pliable.

Stage 3

Once the plastic becomes pliable the former is raised into the plastic, sealed and the air sucked out. Atmospheric pressure then forces the plastic around the former, thus taking on its shape.

Stage 4

The thermoplastic cools and is removed from the machine.

The advantage of vacuum forming is that it is more cost effective than other manufacturing processes. The former in vacuum forming is cheaper to produce than the mould in injection moulding, which means that this process is, in particular, more cost effective on small or medium production runs.

FACTFILE:

Process	Advantages	Disadvantages	Polymers used	Applications
Blow moulding	• Intricate shapes can be formed • Can produce hollow shapes with thin walls to reduce weight and material costs • Ideal for mass production: low unit cost for each moulding	• High initial set-up costs as mould expensive to develop and produce	• HDPE • LDPE • PET • PP • PS • PVC	• Plastic bottles and containers of all sizes and shapes, e.g. fizzy drinks bottles and shampoo bottles
Injection moulding	• Ideal for mass production: low unit cost for each moulding for high volumes • Precision moulding: high-quality surface finish or texture can be added to the mould	• High initial set-up costs as mould expensive to develop and produce	• ABS • HDPE • Nylon • PP • PS	• Casings for electronic products, containers for storage and packaging
Vacuum forming	• Ideal for batch production: inexpensive • Relatively easy to make moulds that can be modified	• Mould needs to be accurate to prevent webbing from occurring • Large amounts of waste material produced	• Acrylic • HIPS • PVC	• Yoghurt pots, blister packs • Insides of fridges etc.

Rotational moulding

The rotational moulding process is a plastic forming process that produces hollow one-piece components. A measured amount of powdered polymer is loaded into a mould. Heat is applied to the mould and at the same time the mould is rotated in a tumbling action until all the polymer has melted and has stuck to the mould wall. The mould is then cooled either by water or air and the component is removed.

FACTFILE:

Applications, advantages and disadvantages of rotational moulding		
Advantages	**Disadvantages**	**Applications**
• Easy to produce large products • End product has no seams • Wall of product is uniform thickness • Corners of product are stress free • Colour integrally part of product • Metal inserts can be included in the moulding • Products are near-net-shape and rarely need further finishing.	• Lower volume production • Materials available are limited • Can be labour intensive when compared with injection moulding • The long cycle times usually limit economic batch sizes to between 500 and 10,000.	• Buckets • Plastic footballs • Dustbins • Oil drums • Storage tanks • Traffic cones

WEBLINKS:

www.bpf.co.uk/

Website of the British Plastics Federation with a lot of useful information

Manufacturing techniques for mass production

Jigs

When manufacturing a product, there are sometimes instances when a particular process or construction needs to be repeated a number of times. This might include producing a hole in exactly the same position on a number of pieces of metal. To mark out exactly the same place on many similar components could take a long time and there would be a chance that, if not undertaken carefully, inaccuracies would creep in. In order to overcome this, a jig could be used.

A jig is a device designed to hold a component in place. For example, a jig could be clamped to a drilling machine and work slotted into it to enable holes to be drilled. Because the jig is clamped to the drilling machine, when the component is put into the jig, the drill is always lined up exactly above the location of the hole. This fulfils two roles – firstly, the hole is accurately drilled and secondly the production process is speeded up.

Jigs can be used in many situations. They can be used in cutting processes, to ensure that components are always the same size and they can be used in milling and routing. Jigs are particularly useful when a number of components are required such as in a furniture factory where many table legs are needed with identical holes or slots in the same place on each component. Using jigs means that the drilling can be undertaken very efficiently.

Patterns

Patterns are replicas of products, usually made from wood that are produced for casting. Pattern making is a highly skilled job as many factors need to be considered to ensure that the cast item is produced to the correct size with no faults or blemishes. Patterns must be carefully designed to ensure that they can be removed from the sand successfully. They should allow for contraction of the metal and should take into consideration cores and holes that might be required on the finished product. (See casting pages 43–4 for full details of the process.)

Formers

Formers are used in processes such as vacuum forming and laminating. Formers are shapes around which material can be bent or formed in order to create, in some instances, quite complex components. An example is vacuum forming, where a thermoplastic is heated and atmospheric pressure forces the plastic around the former. When the plastic is cooled, it is removed from the former and, because it is a thermoplastic, it retains the shape of the former. Other examples include using formers in lamination to produce different shapes.

There are advantages and disadvantages to using formers. They can be used to create innovative work, for example if a designer wishes to create a piece of plywood furniture with sweeping curves, then the material can be shaped using formers. A disadvantage of using formers for complex constructions is that they can be cumbersome and difficult to handle. When used in vacuum forming, the design of the former becomes critical as consideration must be taken of the extraction of the air and the way in which the heated thermoplastic shapes around the former itself.

Moulds

Moulds are the 'negative' in processes such as casting, injection moulding and blow moulding. The mould is where the molten metal or molten plastic is either poured or forced in order to make the desired shape. In casting the mould might be made from sand or hardened steel. In sand casting, the moulds are used once and then the sand is either reconstituted or, if Petrabond® is used, discarded (see pages 43–4 for details of sand casting). In the case of die casting or injection moulding the mould is used over and over again. This means that the design and manufacture of the mould is extremely important. If mistakes are made it can put back the production process and thus cost a company a great deal of money. The designing and production of moulds for die casting and injection moulding is complex and requires highly skilled engineers.

THINK ABOUT THIS!

Jigs are a useful tool when manufacturing a number of items that have similar characteristics such as holes or slots. Think about your project work and design a simple jig that could be used on one area of your project. If you are making a table, you may well require four identical legs with holes drilled in the same place on each leg. It may save time to construct a jig instead of marking out and drilling each hole individually.

Go + No go gauges

Joining techniques

Nuts, bolts and washers

Nuts and bolts are a way of fixing materials together. Bolts come in a variety of sizes in terms of both length and diameter. The majority have hexagonal heads that can be easily tightened and undone using spanners, although some do have square heads. It is important that the thread form of the nut and bolt are compatible. If this is not the case, the bolt will not screw into the nut. There are many types of thread form, but in general it is the metric thread that is most commonly used today. Once they are joined together, the nut and bolt produce a very strong mechanical joint. However, the important thing about nuts and bolts is that they form a temporary joint. They can be undone and reused an infinite number of times.

Washers are thin disks, usually of metal, but can be of plastic or other materials, with a hole through the centre that are often used in conjunction with nuts and bolts. A washer's main function is to help distribute the load when tightening a nut. They can also be used as spacers, or as methods of sealing in situations where gases or liquids might be used and specialist washers can lock nuts into position.

Rivets

Rivets are used to make permanent joints in metal. The two main types of rivet are solid and pop. Solid rivets come in two forms, snap (round head) and countersunk head. Pop rivets come in only one form but in various sizes.

Snap riveting

Snap rivets have round heads and special riveting tools called the 'snap' and 'set' are required for the riveting process. The snap, which has a domed indentation that matches the round head of the rivet, is used to support the round head of the rivet during the process. The snap is secured in a vice and the rivet is rested in it. The set, which has a cylindrical hole to match the shaft of the rivet, is used to hammer the rivet down into its correct position.

Stage 1	• Mark out the positions • Drill the holes through the material in desired positions
Stage 2	• Clean off burrs and waste material
Stage 3	• Place the rivet through the holes and support the round head of the rivet in the snap which has been secured in the machine vice
Stage 4	• Check the length of the shaft of the rivet • There should be 1½ times the diameter above the work to be joined
Stage 5	• Use the ball-pein of the hammer to shape the rivet into a round head
Stage 6	• Finish off using the snap and hammer to make the rivet a smooth round shape

Table 2.16 *Stages of snap riveting*

Countersunk head rivets

Countersunk head rivets are used where it is important that the head of the rivet is flush with the surface of the materials being joined.

Stage 1	• Drill the holes through the material in the desired positions on both of the outside faces of the join
Stage 2	• Remove any burrs and waste material
Stage 3	• Place the rivets into the hole and press the metal to be joined together
Stage 4	• Place the countersunk head on flat metal surface
Stage 5	• Hit the rivet with the flat face of the hammer to swell the rivet
Stage 6	• Use the ball-pein of the hammer to force the rivet into the countersink
Stage 7	• Finish off with the flat face of the hammer and then smooth off with a file

Table 2.17 *Stages of countersunk riveting*

When a countersunk rivet has been used the work should be filed flush so that it is virtually impossible to see the rivet.

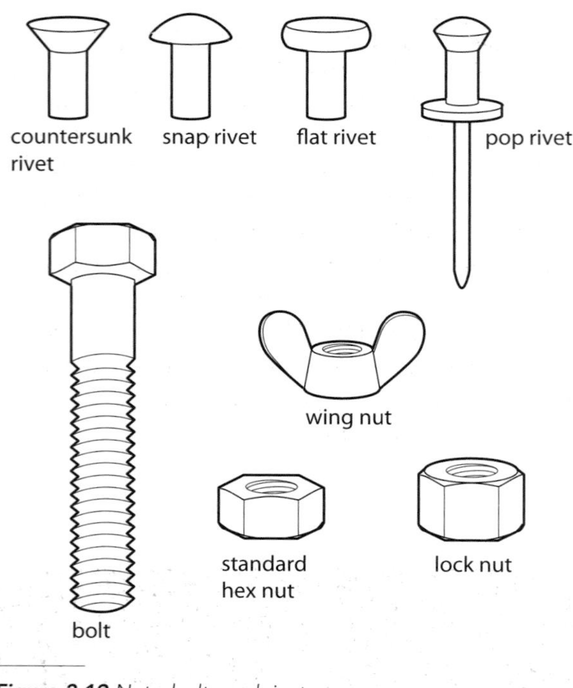

Figure 2.19 *Nuts, bolts and rivets*

Pop rivets

With solid riveting, both sides of the work need to be accessible. There are occasions however, when working on both sides of the material is either impractical or impossible; in such cases, pop riveting is used. To pop rivet, a pop rivet gun is used. The pop rivet itself is hollow and has running through its centre a piece of wire called a mandrel. The rivet is placed into the holes in the pieces of metal to be joined and the rivet gun is used to pull the mandrel through the centre of the rivet. As the mandrel is being pulled, the end of the inner side of the rivet begins to expand, pulling the two sheets together. Eventually, when the rivet has been pulled to its limit, the mandrel snaps and leaves the rivet securely in position.

Welding

Welding is another method of joining metal, this time using heat. A weld is a 'fusion' joint. This means that the two pieces of metal are literally melted together. At the end of the process the two parts that have been joined literally become the same component.

There are two types of welding – oxyacetylene welding and arc welding. With oxyacetylene welding, the two gases oxygen and acetylene are mixed in a gas torch, ignited and used as the source of heat. A filler rod of the same material is often used to complete the joint. At the end of the process the joint should be as strong as the parent metal. Some metals are easier to weld than others. An example is aluminium which is difficult to weld as it has a very low melting point and when heated oxidises very quickly. The most commonly welded metal is steel.

Metal Inert Gas (MIG) welding is a form of electric arc welding. MIG welding involves a process where, using an electric current, an arc is struck between the work and an electrode and this is used as the heat source. The filler is in the form of thin wire and as the welding progresses the wire gradually is fed into the joint and is used up. During the operation a flow of inert gas, usually argon, is made to flow over the area being joined. The inert gas acts as an envelope that keeps oxygen away from the joint. This prevents oxidation on the joint and therefore helps to make the weld sound. This process is ideal for use on materials that cannot be gas welded, for example, aluminium.

Brazing

Brazing is a method of joining mild steel to mild steel using heat and a second material called spelter. Spelter is an alloy of copper (65 per cent) and zinc (35 per cent) and has a melting point of 875°C.

FACTFILE:

Name	Advantages	Disadvantages	Applications
Nuts and bolts	• Can be applied and removed an infinite number of times	• Prone to vibration	• Engineering situations where joint needs to be undone
Snap rivets	• Good, strong joint	• Access to both sides of materials being joined required	• Engineering situations where permanent joint is required
Pop rivets	• Can be applied from one side	• Relatively weak when compared to snap rivets	• Joining thin sheet metal

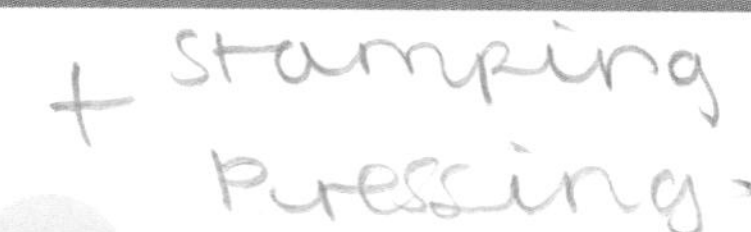

In brazing, the preparation of the two pieces of metal prior to joining them is very important. The joint area should be cleaned up with emery cloth. Then a flux should be applied to the area where the join will be. A flux serves two purposes. Firstly, when the steel is heated up, there is a reaction between the oxygen in the air and the metal. This causes oxidation. When oxidation occurs a brazed join will not work because the molten spelter will not flow into the join. The flux prevents oxidation because it forms a barrier between the work and the atmosphere. Secondly, a flux breaks down the surface tension on the molten spelter and allows it to flow between the two pieces to be joined.

Brazing is usually undertaken on the brazing hearth. The work, with the spelter and flux applied, should be carefully arranged on refractory bricks and then heated up using a brazing torch. When the work reaches 875°C the spelter will melt and flow into the join. Once completed the work is allowed to cool down and then removed from the hearth. Finally, the joint will need to be cleaned up and all traces of the flux removed.

Figure 2.20 *Gas welding in industrial situations.*

Hard soldering

Hard soldering is a process that is very similar to brazing. It can also be referred to as silver soldering. It is most commonly used when joining such materials as copper or in jewellery-making. Like brazing, hard soldering is a capillary joint. It too requires a flux to allow the solder to flow and to help prevent oxidation. The main difference between brazing and hard soldering is the temperature at which the hard solder melts. The melting temperature of hard solder ranges from 625°C to 800°C. The fact that different grades of hard solder have different melting points can prove useful when attempting to join a number of components onto a single piece of work.

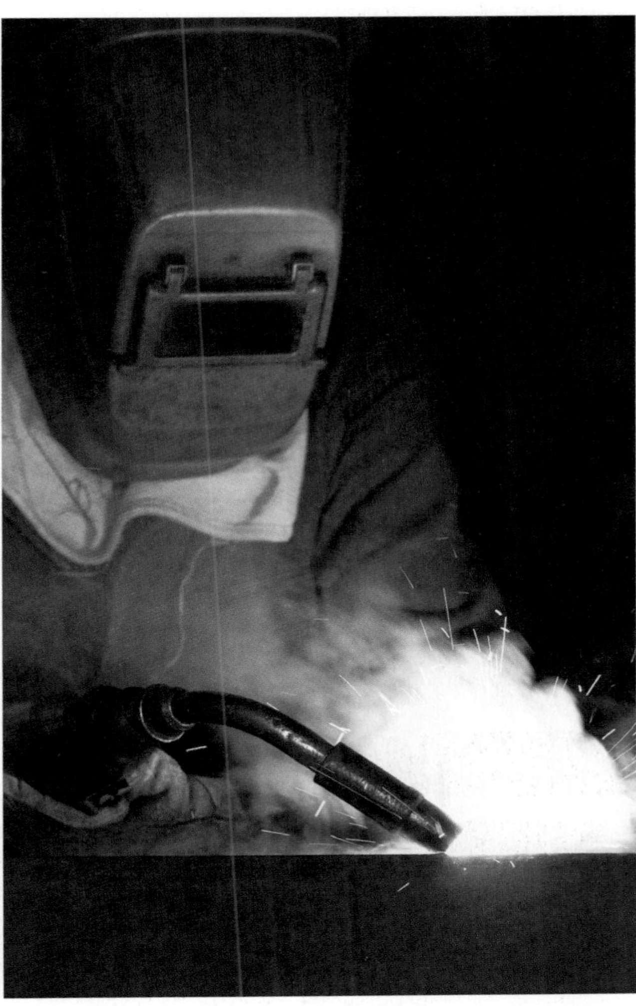

Figure 2.21 *MIG welding in industrial situations.*

FACTFILE:

Joining process	Advantages	Disadvantages	Applications
Welding	• Weld is as strong as parent metal	• Can be safety issues with both gas welding and MIG welding	• General engineering • Automotive engineering
Brazing	• Good general purpose joint • Can be undertaken with little training	• Care needed when undertaking braze to ensure joint is clean and oxides do not form creating weak joint	• General engineering
Hard soldering	• Relatively low temperatures required • Comes with different melting points to enable complex decorative pieces to be produced	• Weak joint does not withstand stress	• Decorative metalworking

Chemical joints

Tensol cement

Tensol cement is used to join thermoplastics. In some ways a join using Tensol cement is very similar to welding a joint. In fact it could be described as a 'chemical weld'. Tensol is a solvent (dichloromethane) and when it is applied to a thermoplastic it will literally melt the material. The Tensol cement should be applied to both surfaces of the thermoplastic to be joined and then the component parts should be pushed together. The join then needs to be held carefully together. Ideally the join should be clamped together for up to 24 hours. The result is a permanent joint. It should be stressed that as Tensol cement is a solvent it gives off fumes and should therefore be used with caution. It is regarded as a hazardous substance.

Polystyrene cement

Polystyrene cement works in a similar fashion to Tensol cement. It too works as a chemical weld. The cement contains a solvent that melts the surface of the plastics being joined. This allows the two component parts to fuse together. As the solvent evaporates the strength returns to the plastic. Like Tensol cement, any parts being joined need to be held together until set.

It should be remembered that only thermoplastics can be joined with Tensol cement or polystyrene cement. Thermosetting plastics will not melt and therefore cannot be joined with a chemical weld.

Adhesives

Polyvinyl acetate

Polyvinyl acetate (PVA) is probably the most widely used wood glue. It comes as a thick white viscous liquid and is very useful for gluing together all different kinds of wood. It is water soluble which means that it can have advantages over other glues in some situations. Before applying PVA a number of issues do need to be addressed. The wood surfaces that are to be joined together must be clean and

FACTFILE:

Type of cement	Advantages	Disadvantages	Applications
Tensol cement	• Gives excellent bond when joining acrylic	• It must be applied with care to avoid damage to surface of acrylic • Gives off strong fumes	• Joining acrylic
Polystyrene cement	• Gives excellent bond when joining two pieces of polystyrene	• It must be applied with care to avoid melting polystyrene	• Joining polystyrene • Used on Airfix® kits

match each other with no gaps. If the surfaces are not in contact then the gluing up will not work. PVA needs to be applied sparingly. Although it is water soluble if it gets on pieces of work by mistake, it will stain the wood. It takes approximately 24 hours to harden, so once it has been applied to the work being glued, the work needs to be clamped together.

The most common clamps used in woodworking are G clamps and sash clamps. As work is being glued up it is vital to check all the measurements and that the work is square where it should be. Another thing to remember when clamping up is to always put a piece of scrap wood between the clamp and the work. This prevents damage to the work by distributing the load as the clamp is tightened. Once the PVA has become hard it is impossible to undo the joint and re-glue it.

Epoxy-resin

Epoxy-resin is a thermosetting plastic resin that cures, or goes hard, when mixed with a catalyst or hardener. Once the two materials, resin and catalyst come in contact with each other, a chemical reaction begins and the resin begins to go hard. This reaction is not reversible so once it starts it cannot be stopped.

Epoxy-resins are very useful because they can be used on almost any material in any situation and the join they produce is very strong. The most common type of epoxy-resin available in school is Araldite®.

To use an epoxy-resin, equal quantities of resin and catalyst are squeezed out of their tubes onto a piece of scrap wood or metal and then mixed together. Once they are completely mixed, the resultant paste can be applied to the surfaces being joined. It is advisable to wear latex gloves when mixing and applying the resin as contact with the skin in inadvisable. The process of curing, or becoming hard, can take a number of hours and ideally, during this time, pressure needs to be applied to ensure that the material remains in the required position.

Contact adhesives

Contact adhesives are made up of natural rubber and polychloroprene (Neoprene). They are very useful in situations where materials like laminates need to be glued to flat surfaces. An example of this is when they are used to glue plastic laminates such as Formica® onto kitchen worktops.

As with all gluing processes, the first thing to remember when using contact adhesives is to make sure that the surfaces being bonded are perfectly clean and free from dust. Contact adhesives need to be applied to both surfaces being joined. The adhesive must be applied thinly on those surfaces and then left to dry for a few minutes. You can tell if it is dry enough by touching the adhesive – if it is ready it will feel dry to the touch.

Once the adhesive is dry the two surfaces can be pushed together. When the two elements come into contact with each other the bonding occurs almost immediately. This instantaneous bonding is useful as no clamping is required though, it does mean that once contact is made between the components it is impossible to adjust the join. Contact adhesive must be used in well-ventilated areas as it gives off toxic fumes.

Hot melt glue

Hot melt glue is the glue that is used in glue guns. The glue itself is a thermoplastic adhesive and, like an ordinary thermoplastic, melts when it is heated up. The glue itself is in the form of cylinders that come in various diameters and various lengths to enable them to fit into different glue guns. The glue gun contains an electrical heating element that, when switched on, heats up and melts the glue. The glue gun has a trigger arrangement that when squeezed forces the molten glue through the nozzle in the barrel of the glue gun. This type of adhesive is very useful for small model making tasks. Once the glue has been applied it is tacky but once it begins to cool down it produces a very sound bond.

Like many things in Resistant Material Technology, there are safety issues that need to be considered when using a glue gun. The molten glue is very hot and if allowed to come in contact with the skin it will stick to the skin and can cause blistering.

Figure 2.22 *Glue gun*

+ Cyanoacrylate (Super glue)

Type of glue	Advantages	Disadvantages	Applications
Polyvinyl acetate (PVA)	• Produces strong bond when joining wood • Can be used on cards and papers	• Can take up to 24 hours to bond wood • Requires clamps to hold work in position whilst hardening	• Wood joints
Epoxy-resin	• Very strong • Waterproof • Heat and chemical resistant	• Takes time to cure and harden • Requires two elements: resin and catalyst	• Aircraft • Boats • Golf clubs • Skis
Contact adhesives	• Instant sticking	• Cannot be adjusted during joining process	• Applying Formica® and veneers
Hot melt glue	• Good bond when used in model making • Relatively quick to harden and bond	• Safety issues: glue gun can be very hot and can burn	• Model making

Material removal

Cutting

Saws

As you work through your project work, you will need to make decisions and one of the issues that is important to think about is choosing the correct methods of cutting material.

Different saws are used for various tasks. You are required to use saws for metalworking and saws for woodworking tasks.

Metalworking saws

The principal metalworking saw is the hacksaw. It is a frame saw and the frame holds a separate replaceable blade under tension. The blades come in a variety of lengths and the frame is adjustable to accommodate those various sizes. A hacksaw blade is classified by the number of teeth per 25mm, and the most common size used for general purpose cutting is a blade that has 25 teeth per 25mm. When replacing a hacksaw blade care should be taken to ensure that the teeth face forwards. Positioned this way, the saw will cut on the forward stroke. A smaller version of the hacksaw is the 'junior' hacksaw. This is used for more intricate cutting.

Woodworking saws

In woodworking, different saws are used for various tasks. Woodworking saws come in a number of shapes and sizes and it is important that you know which saw is used for which job. If you wish to cut along the grain, you need a large saw with big teeth that can remove the waste material such as a rip saw. To cut across the grain, you would use a cross-cut saw. These have slightly smaller teeth. For small accurate work like cutting out joints, a back saw would be used. The tenon saw is an example of a back saw. With back saws, the blade has a strengthening piece of steel or brass along the top of the blade to keep it rigid whilst cutting. The size of a saw is specified in teeth per 25mm. The general rule is that saws with smaller teeth are used for hardwoods and saws with large

Name	Teeth per 25mm	Uses	Notes
Hacksaw	25	• General purpose metal saw	• Hacksaw blade is held in frame • Blade must be kept in tension • Teeth must always point away from the handle • Can be adjusted through 90°
Junior hacksaw	32	• Small general purpose saw can be used on most metals	• Used for straight cuts on small pieces of work • Blade is held in tension by sprung steel frame

Useful GCSE Knowledge

teeth are used to cut softwoods. It is important that wood saws are maintained properly and their teeth are kept sharp so that they cut efficiently. The first stage in maintaining saws is to ensure that the teeth are sharp. They are sharpened using a 'saw file'. These files are triangular in shape. Sharpening a saw requires considerable skill and can take some time.

The teeth on a saw are always arranged so that they are bent slightly, alternately to the left and to the right. This is done to enable a slightly larger groove than the width of the saw to be cut. This gives the saw blade clearance as it cuts and prevents it getting stuck or creating too much friction. The angle on the teeth is referred to as the 'set'. Once the teeth have been sharpened, the set is checked and adjusted using a tool called a 'saw set'. For intricate work or sawing a curved line, a coping saw is used. Coping saws are frame saws. The blade is kept in tension with a sprung steel frame.

FACTFILE:

Name	Points per 25mm	Length	Uses
Rip saw	5	700mm	• Sawing along the grain of large pieces of wood
Cross-cut saw	7	600mm	• Sawing across the grain of large pieces of timber
Tenon saw	14	Various sizes from 250mm to 350mm	• Producing joints and work on small pieces of wood

Shearing

Sawing is not the only way of cutting materials. When cutting sheet metal for example, it is not always possible to use a hacksaw because the frame gets in the way. In these circumstances, it is sometimes more appropriate to use bench shears or tin-snips. Bench shears are mounted on the work bench and can be used to cut very large sheets of material such as aluminium, tinplate or even steel. Tin-snips are hand-held and used for smaller cutting tasks. The action of bench shears and tin-snips is one of pressing or shearing the material. Bench shears and tin-snips do not have sharpened blades.

Abrading

Abrading is the wearing down or rubbing away of material. In terms of Resistant Materials Technology this means filing, grinding and sanding.

Filing

Filing is a method of removing metal. Various files are used for different tasks. They come in a wide variety of shapes and sizes with different lengths and different cross sections. The most common file is the hand file. This has a rectangular cross section and has parallel sides. Only two sides and one edge have teeth. The second edge is referred to as a safe edge and can be very useful if you are filing up to an edge on a piece of work. Other files include round, half-round, square and triangular cross sections. There are also other specialist files; warding files are very small and thin and designed for filing narrow slots and needle files are very small files that come in a variety of shapes for specialist work and dreadnought files which are used on soft metals such as aluminium. The teeth on files can vary in size; the larger the teeth the more metal can be removed. It is usual when working on a project that you start with a file with large teeth and finish off with a much smoother file.

Most files are used on metal. However, one type of file, the rasp, is for using on wood. Unlike metalworking files, a rasp has fairly large individual teeth. These large teeth are designed so that waste wood does not clog up the tool as it is being used. Files sometimes get clogged up and need to be cleaned or pinned. To do this a file card is used to brush out any pieces of material that have become stuck in the teeth. One safety point to remember when using a file is that it should never be used without a handle. The handle is fixed onto the file by a sharpened point called a tang, which if removed is a potential hazard.

Grinding

Grinding is a highly specialised operation and should only be undertaken by trained technicians. **Under no circumstances should students attempt to undertake grinding.** There are three types of grinding: offhand, disk and surface.

Offhand grinding is where a machine which is either mounted on a bench or on the floor powers a grinding wheel. These are often found in school workshops and are used for sharpening tools such as chisels. Different types of grinding wheel are used for different purposes and the users of such machines need to be familiar with the types of wheel and what material they can be used on. They must also be trained in the correct procedure for mounting grinding wheels.

A disk grinder is held in the hand and is used as a portable unit and these are usually used as a finishing off tool, perhaps for grinding down a weld.

A surface grinder is a machine that is able to precision grind component parts to very fine tolerances.

Setting up and operating surface grinders requires highly skilled technicians. Grinding can be extremely dangerous. When grinding, hot metal particles fly off the work and there is also a risk that, if not correctly undertaken, the grinding wheel itself could burst. Anyone grinding should have received the correct training and must wear appropriate protective clothing.

Sanding

Sandpaper is a general term that is used to describe paper that has abrasive grit glued onto a backing sheet. The correct term to use is actually 'glasspaper'. The grit on these papers can be glass, aluminium oxide, silicon carbide or garnet and each of these grits has a different characteristic and can be used for a different task. Glass-paper is classified by number and ranges from 40 the coarsest (roughest) to 400 (smoothest) which is incredibly fine. Glasspapers are also graded from coarse to very fine. The grit size and the grade of the glasspaper are usually printed on the reverse side.

FACTFILE:

Glasspaper grades	
Grit size	Grade
40–60	Coarse
80–100	Medium coarse
120–150	Medium
180–220	Fine
240 and greater	Very fine

It is important to remember that when glasspapering the glasspaper should be totally flat on the wood. The best way to achieve this is to wrap the glasspaper around a cork block and use the block to keep the sanding flat. When rubbing down work it is best to use various grades of paper, starting with a coarse grade and then gradually working down through the grades until finishing off with a fine paper.

Emery cloth

The equivalent abrasive in metalworking is emery cloth. Like glasspaper, emery cloth comes in various grades and when using emery the most abrasive should be used first, going down to the finest to finish off.

THINK ABOUT THIS!

Cutting and glasspapering are a very important part of the preparation of a joint before gluing up. Write down the stages of producing a dovetail joint in wood, naming all the processes and all the tools that are required.

Heat treatment

Work hardening

Work hardening occurs when a non-ferrous metal is cold worked. This means it is bent, hit or shaped over a period of time. As the material is shaped it becomes so hard and brittle that it can eventually fracture. Sometimes this process can be useful such as when making a decorative copper dish, as it will be necessary to soften the metal to enable it to be manipulated into the correct shape. Once the desired shape has been achieved, in order to keep it in that shape, it needs to be hardened. In order to do that, the work would be plannished – that is hit gently with a hammer to work harden the copper to stop the dish bending.

Alternatively in something like an aircraft, work hardening is not a good thing because if the aluminium alloy of an aircraft becomes work hardened the metal will split and the consequences can be catastrophic. Work hardening can also occur during processing where metal is cold rolled or cold drawn.

Annealing

When a piece of metal is work hardened and the material becomes very difficult to bend and shape, it becomes necessary to relieve the stresses that have been built up inside the metal. The process used to relieve the internal stresses within non-ferrous metals and make them workable again is called annealing. Different metals are annealed in different ways but all methods of annealing involve the application of heat.

Annealing aluminium

When annealing aluminium, the biggest problem is that it has a very low melting point and does not change colour. This means that when it is heated up there is a danger of the metal melting.

- Apply soap to aluminium as a temperature indicator.
- Gently heat the aluminium until the soap turns black.
- Using tongs remove the aluminium from the heat source.
- Cool under running cold water.

Annealing copper

- Heat copper until it turns a 'dull red' colour.
- Using tongs remove copper from the heat source.
- Allow copper to cool down in air or quench.
- Place copper in an acid bath to remove oxides.
- Remove acid from work by placing under running cold water.

When copper is heated oxides form on the surface of the metal which, if allowed to remain, can damage and pit the surface of the metal. It is important therefore to remove the oxide layer. In order to remove the oxides from the copper, the material needs to be immersed in a bath containing a solution of sulphuric acid and water (1 part acid to 10 parts water). This process is called pickling. After pickling, the copper is removed from the acid and then carefully washed under running water.

Great care should be taken when annealing and the correct protective clothes, such as aprons, gloves and goggles should be worn. Once metal has been annealed it can once again be worked and manipulated. However, if the metal again becomes work hardened it may well need to be re-annealed. In fact the process may well need to be carried out a number of times during the manufacture of a single item.

Normalising

Normalising is a process that is undertaken on ferrous metals that have become hardened, in order to return them to their original unhardened state. When medium or high carbon steel is heated to its critical temperature, then plunged into water, it will harden. If an attempt is made to cut or drill the material when it is in this state, the hacksaw or drill will soon become blunt. In order to remove the hardness, the steel must be normalised. The steel is heated until it is cherry red, 900°C hot and then allowed to cool down in air. Once the steel has cooled it will have lost some of its hardness and can, once again, be cut or drilled.

One factor that should be remembered during the normalising process is that once the heat source has been removed the metal will lose it 'redness' very quickly. It will return to its original colour. However, the material will still be extremely hot and care must be taken to label the work as it is cooling to avoid injury.

Hardening and tempering

If carbon steel is heated to red heat, 900°C, and then quenched in water, it becomes very hard. At the same time as becoming hard, its brittleness increases. The consequence of this is that the steel is likely to break if it is put under pressure. One example is a screwdriver needs to be hard. It needs to be slotted into a screw and withstand the turning forces that are put on it during tightening a screw. However, if the screwdriver is hard but brittle, as the screw is turned the end of the screwdriver is likely to snap because of its brittleness. In order to remove the brittleness from carbon steel in a product such as a screwdriver, the piece of work will need to be first hardened to allow it to take the pressure of the work, but then tempered to remove the brittleness in the material.

To harden steel, the work is heated gradually until it is red hot. When the steel is red hot the work is removed from the heat source and then plunged into cold water. This action causes the steel to cool rapidly and as a consequence, the steel becomes very hard and brittle. In order to remove the brittleness, the steel is cleaned up using emery cloth until it is shiny. The steel will then need to be tempered. It is then reheated, this time very slowly and carefully. As it is being heated a thin line of oxide will appear on the steel. This line of oxide will change in colour as it gets hotter. It will start as a pale yellow colour, go though dark yellow, brown, purple and then blue. For a screwdriver, the colour of the oxide needs to be blue. When

Oxide colour	Approximate temperature	Component
Pale yellow	230° (hardest)	• Lathe tools • Scribers • Dividers
Dark yellow	250°	• Drills • Taps and dies • Hammer heads
Brown	260°	• Shears • Plane blades • Lathe centre bits
Purple	270°	• Knives • Axes • Woodworking tools
Dark purple	280°	• Saws • Table knives • Cold chisels
Blue	300° (toughest)	• Screwdrivers • Springs • Spanners

Table 2.18 *Tempering oxide colours*

the desired oxide colour is reached the work is removed from the heat and quenched. As the metal is quenched the iron carbide content is frozen at a particular level. Any residual heat would continue to course normalisation if quenching was not carried out. In the case of the screwdriver, when it has been tempered, it will be hard but not brittle so it will be able to be used without danger of snapping. Tempering removes some of the brittleness and replaces it with toughness. Other components are heated to different temperatures.

Conversion and seasoning

As soon as possible after a tree has been felled, it needs to be converted. Conversion is the term used to describe the process of changing the wood in the tree into usable planks of timber. There are two common methods of converting timber, the through and through (slab) method and the quarter sawn method.

Through and through (slab) sawn conversion

The through and through (slab) method is the simplest method of conversion and involves cutting the timber along the complete length of the log. The result is long planks of parallel cut wood. The advantage of this method is that it is seen as a relatively cheap way of converting timber and there is little waste. However, there is a downside. The final planks, because of the arrangement of the annual rings, may be prone to warping and cupping. This method is more often used in the conversion of softwoods.

Quarter sawn conversion

Quarter sawn conversion is a more expensive method of conversion than the through and through method. This is due to the fact that there is far more wastage. The timber is cut in such a way so as to make the annual rings as short as possible. This reduces the risk of the wood twisting or cupping (see page 63). As a consequence of this quarter sawn converted timber is usually regarded as higher quality than through and through converted timber.

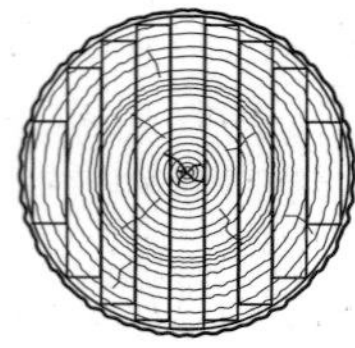

a) slab sawn

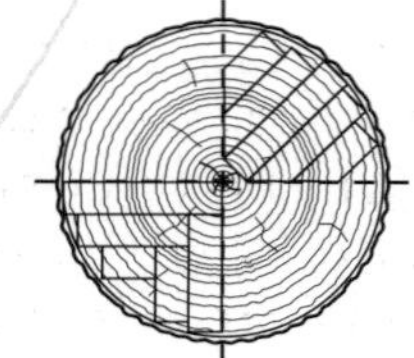

b) radial 'quarter' sawn

Figure 2.23 *Slab and radial sawn logs*

Seasoning

Seasoning is the removal of excess moisture from timber after the timber has been felled and converted into usable planks. There are two different types of seasoning: natural seasoning and kiln seasoning.

Natural seasoning

Natural seasoning is where the timber is allowed to dry out at its own pace in air. In natural seasoning the timber, once cut into planks, is stacked. Sticks are placed between the layers of planks to separate them and to allow air to circulate around the wood. The stack is raised off the ground and has a roof put over it to protect it from the worst of the weather. The ends of the planks are painted to prevent the ends drying out faster than the rest of the timber.

The stack is then left to allow the timber to dry out slowly. This is the cheapest method of seasoning, but it does have some disadvantages. As the process is undertaken in totally uncontrolled conditions, there is no way of establishing the water content of the timber and there could be some inconsistency with the final product. Another disadvantage is that it can take up to 5 years for the wood to season and dry out ready for use. As a general rule, wood is seasoned for one year for every 25mm of thickness of timber.

Kiln seasoning

Kiln seasoning is where timber is dried out 'artificially' in a kiln. The planks of wood are stacked on a trolley with small pieces of wood to separate the layers. The trolley is then rolled into the kiln and the kiln is then sealed. Once the wood is in the kiln, steam is pumped into the chamber. This has two effects. Firstly, it kills off any bugs and insects that may be in the timber. Secondly, it has the effect of forcing moisture from the cells in the wood, which ensures that the moisture content is the same throughout the timber.

Once this has been done, the steam level is reduced and the temperature gradually increased. This dries out the wood. At the end of the process a flow of almost dry air is pumped into the kiln. This process ensures that the timber is seasoned in a completely controlled way. When the timber is withdrawn from the kiln the moisture content is usually below 12 per cent, which makes it suitable for most applications. The advantage of kiln seasoning is that it is far quicker than drying out wood naturally and because the moisture content can be controlled, the resulting timber is very stable. The disadvantage of kiln seasoning is economic. It is far more expensive than natural seasoning.

Faults in wood

As wood is a natural material it can be prone to problems. Problems can result from a number of factors. It might be issues with seasoning where uneven drying can cause faults, or it might be as a result of biological or fungal attack. It is always a good idea to check timber thoroughly before using it. Poor-quality timber can affect the final outcome of what otherwise would have been a high-quality product.

The principal faults in wood are:

- **Cupping**
  A piece of timber is said to be cupped when it is bent across the grain. When a plank of wood is viewed from the end the annual rings can be clearly seen. Some rings are longer than others. The longest rings are those that are furthest away from the centre of the tree. The varying length of annual rings causes the wood to dry out at varying speeds. This varying rate of drying causes different parts of the wood to shrink at different rates. The result of this difference in drying rates causes the wood to cup.
- **Twisting**
  If timber is used that comes from the centre of the tree there is sometimes a danger that the wood has a spiral grain. This spiral grain can cause the timber to twist as it dries out. A piece of wood that is twisted resembles a propeller and is usually unusable.
- **Splitting**
  Splitting is when the grain of the timber separates. This usually occurs at either end of a plank of wood. The only thing to do when this happens is to cut off the affected ends of the plank.
- **Knots**
  The knots in timber are where branches were attached to the tree. The tree has continued to grow and wood has covered the point where the branch was. Knots can cause a problem. Sometimes they become loose and drop out. This can be extremely dangerous if the wood is being machined. It is therefore advisable to check for knots when selecting material and avoid timber that has them.

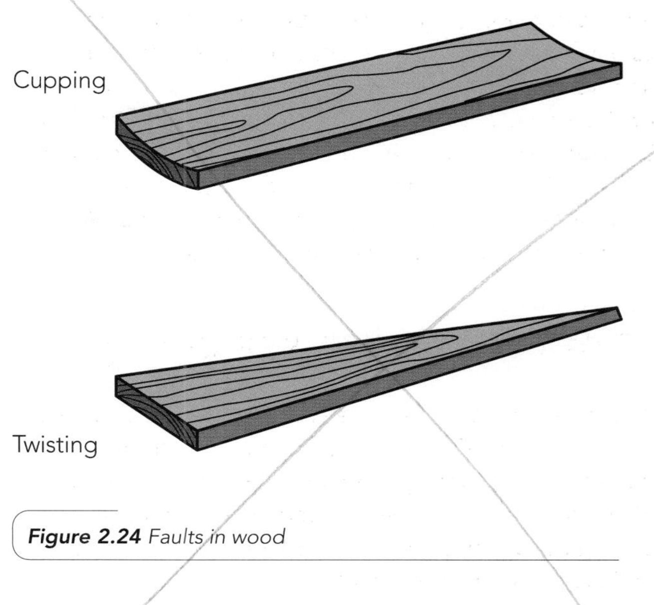

Figure 2.24 *Faults in wood*

Computer-aided design

2D design to create and modify designs

There is a growing trend to move away from more traditional forms of designing, using pencil and paper, to computer-aided design (CAD). There are a number of advantages of using CAD systems. The first advantage is that by using the computer, any modifications to a 2D design can be quickly made and results can be instantly seen on screen. Various components can be designed and drawn and then those components can be 'joined together' on screen to enable in-depth design and development to be undertaken.

Once designs have been produced, elements such as dimensioning, or viewing objects in various ways such as in orthographic projection or isomeric projection, can be easily undertaken. Designs and drawings can be easily stored electronically and transmitted electronically to anywhere in the world. There are some disadvantages. The software for highly complex CAD systems can be expensive and, in many situations, operators of such systems require extensive training before they can understand and operate the system. However, CAD systems are fast becoming the favoured way that design is undertaken in the majority of companies.

3D modelling for creating 'virtual' products

A 3D image can give a more realistic impression than a 2D image. Therefore, many designers will construct new products on screen that, with skill, can easily be modified and manipulated. Product design teams can significantly decrease the time taken to design and develop a new product with 3D modelling, saving development costs and reducing time to market. These virtual products can be tested and evaluated without actually being manufactured and design data can be directly output to a computer-aided manufacture (CAM) system for modelling prototypes.

Name	Advantages	Disadvantages
2D designing and modification	• Highly accurate drawings • Easy to edit drawings • Easy to store drawings	• High cost of set-up • Files can be lost in system if not backed up • High-quality training required
3D virtual modelling	• Can clearly see around an object 360°	• High set-up costs

Table 2.19 *Advantages and disadvantages of 2D and 3D modelling*

Modelling and prototyping

Block modelling

Block modelling is a very useful tool for helping to determine the shape, dimension and surface details by constructing an accurate representation of a final product. Block models have no working or moving parts and are thus concept models of the product that is being, or has been, designed. They are manufactured from a variety of materials, for instance ABS, acrylic and Styrofoam® and have a high finish which makes them look like the real thing. They are often used in advertising and in photographs in brochures.

Block models are produced using cells and sub-cells. By using smaller and smaller cells, more detail can be created and a good representation of a product produced. By using this cell approach, block models can be produced with a very high level of detail. One advantage of using such systems is that, because they are computer generated, as well as the solid block model, elements such as cross sections or plans and elevations can be viewed on the monitor whilst looking at the solid model.

The models can be extremely useful in determining the ergonomic factors of many products. By constructing a number of block models of various shapes and sizes it is possible for designers to literally 'get a feel' for the product. It will soon become apparent in 3D form the designs that are aesthetically pleasing or 'user friendly' and that are worth developing – something that 2D images struggle to achieve.

As with many computer-aided design and manufacturing systems, the main disadvantage is the set-up cost.

Rapid prototyping using CAM/CAD

The need for manufacturing industries to cut down on the time and costs involved in developing a new product has led to the development of rapid prototyping (RPT). This involves the creation of 3D objects using laser technology to solidify liquid plastic polymers, or resins, in a process called stereolithography. Using specialist software applications on a stereolithography machine, 2D CAD drawings are converted to 3D models.

The process is based on the computer slicing the 3D image into hundreds of very thin layers (typically 0.125–0.75mm thick) and transferring the data from each layer to the laser. The laser draws the first layer of the shape on to the surface of the resin, which causes it to solidify. The layer is supported on a platform that moves down, enabling the next layer to be drawn. This process of drawing, solidifying and moving down quickly builds up one layer on top of another until the final 3D object is achieved. Most companies do not have this technology. Instead, they use RPT services from specialist companies. Stereolithography prototypes can typically be delivered within 3–5 days of the client's design data being received, therefore saving both time and development costs.

Advantages	Disadvantages	Applications
• Saves time and money • Client able to see the product throughout the design process • Changes can be quickly incorporated	• Because a rapid prototype is not the real product there is a danger that elements of the design can be missed	• Concept modelling • Solid 3D models • Architectural models

Table 2.20 *Rapid prototyping*

Computer-aided manufacture

Computer-aided manufacture (CAM) is a process that converts drawings that have been produced using a CAD system into actual products. CAM can involve a range of machines, but the most common are lathes, routers, milling machines and laser cutters. These machines are controlled by a computer that guides the machine through an entire manufacturing process – computer numerical control (CNC).

There are many advantages to CAM. The cost of setting up CAM systems can be high, but once the initial costs are made and the system is programmed the advantages are considerable. The machines are able to work 24 hours a day with little human intervention. They ensure that the products they produce are consistently accurate. They are reliable and, because they can be reprogrammed, are very flexible.

CNC machine	Materials cut	Applications
Lathes	• Metals	• Produce accurate and complex shapes that would be difficult to produce on hand operated lathes
Routers	• Wood • Foam	• Similar to milling machine but using softer materials • Used for block modelling • Can cut 2D and 3D shapes
Milling machines	• Metals	• Produce flat and complex curved shapes • In industry work to fine tolerances
Laser cutters	• Plastic • Wood • Card	• Cuts with intense beam of light • Material does not require clamping • Can cut both simple and complex shapes

Table 2.21 *Computer-aided machining*

Quality

Getting started

What actually is quality? A friend might think that a certain product has quality but you might disagree, so, is it subjective? We can all name a product that we consider to be of a high quality, but what qualities does it have? When designing products, quality refers to a product's ability to satisfy a need and, more importantly, its fitness-for-purpose. In manufacturing terms it is producing a product that is the best it can be. That is, fully functional and free from defects.

Quality assurance systems and quality control in production

Quality assurance

Quality assurance (QA) systems are the planned activities used by the manufacturer to monitor the quality of a product from its design and development stage, through its manufacture and to its end use, including the degree of customer satisfaction. In other words, QA is an entire process designed to ensure that the product fulfils all of its requirements.

In the first instance, QA ensures a product is fit-for-purpose using thorough testing throughout the design and development stage. It includes regulation of the quality of raw materials and components that the manufacturer buys in order to start production. QA systems are used to monitor the quality of components, products and assemblies during production using a series of quality control (QC) checks, tests and inspections. Finally, QA supplies fact-based evidence for quality management systems to inspire external confidence to customers and other stakeholders that a product meets all of their needs and expectations.

THINK ABOUT THIS!

We have all probably bought a product and found that it was not of a good quality. What do you suppose was the problem with it – design, materials, manufacture, assembly or finish? Was there a guarantee with it? Using your product analysis skills developed in Unit 1 (see pages 3–9), outline the quality issues relating to a product that you are familiar with.

Quality control

Quality control (QC) is that part of quality assurance that involves inspection and testing of a product, during or immediately after production.

Inspection

Inspection is used to check that manufactured products have been produced within a specified tolerance. Tolerance is defined as the acceptable amount by which a product's size can differ from that stated in the specification. For example, the diameter of a drinks bottle must be such that it fits in the bottle-filling machinery at the bottling plant and also is able

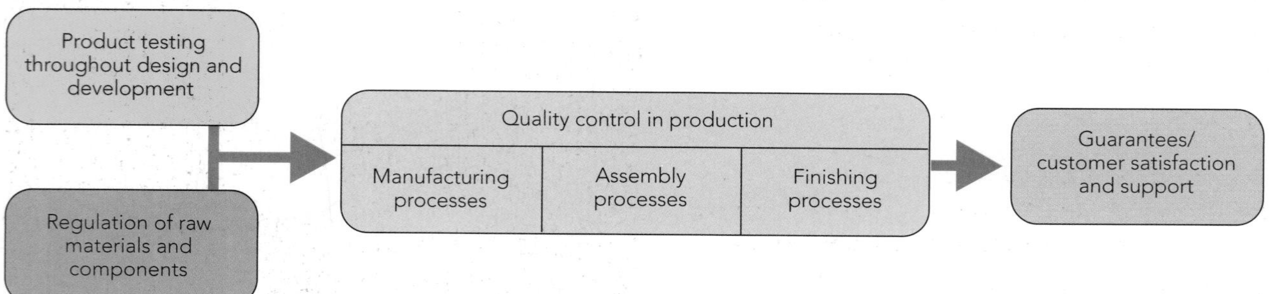

Figure 2.25 *A quality assurance system*

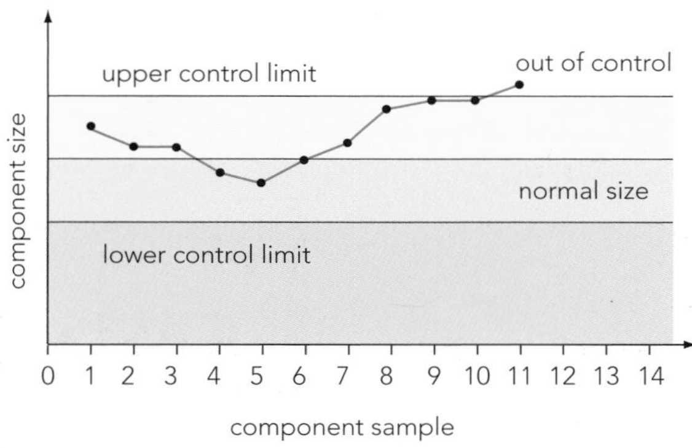

Figure 2.26 *A quality control chart using tolerance*

to contain the required volume of liquid. A 54mm diameter bottle is likely to have a tolerance of +/- 0.8mm. If, when inspected and tested, a bottle measures between 53.2mm and 54.8mm, it will be within the agreed tolerance. Any bottle that falls outside this tolerance is scrapped.

There are three main levels of inspection.

- 100 per cent inspection: where all the units in the lot, or batch, are inspected.
- Normal inspection: using a sampling plan.
- Reduced inspection: using a sampling plan requiring smaller sample sizes than those used in normal inspection. Reduced inspection is used in some inspection systems as an economy measure when the level of submitted quality is 'sufficiently good'.

Computer-aided inspection is possible by using a coordinate-measuring machine (CMM) for dimensional measuring. A CMM is a mechanical system designed to move a measuring probe to determine the coordinates of points on the surface of a workpiece. These machines are used to quickly, and accurately, measure the sizes and positions of features on mechanical parts, with tolerances as small as 0.0025 mm. Laser scanning systems are often used that can determine the coordinates of many thousands of points. This data can then be taken and used to not only check size and position, but also to create a 3D model of the part using a CAD system.

Testing

Testing is concerned with the product's performance. Tests are carried out under laboratory conditions using strict control procedures to ensure that the results obtained are accurate.

There are two main types of tests.

- Non-destructive testing (or failure testing): where the product is tested until it shows signs of failing, for example cracking, to determine how much force is needed to deform it.
- Destructive testing: where the product is destroyed under controlled conditions and monitored to gather valuable research data, for example Euro New Car Assessment Programme (NCAP) for car safety testing.

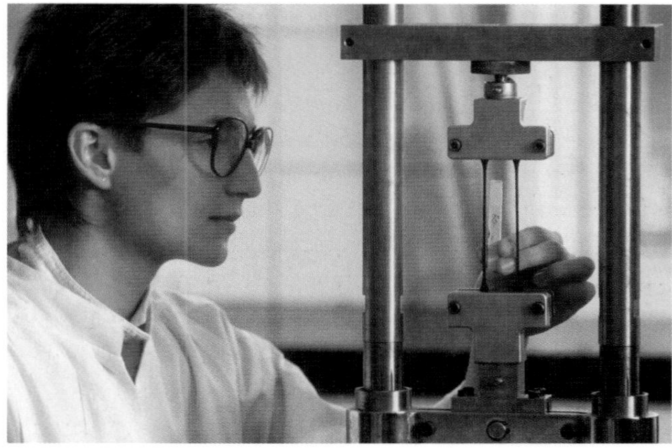

Figure 2.27 *Testing paper tissue for tensile strength*

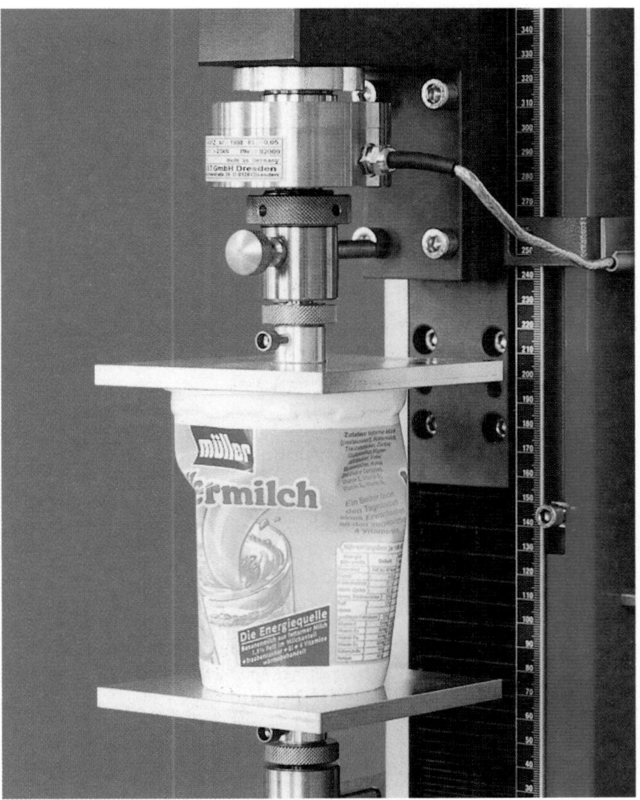

Figure 2.28 *Compressive test of a polymer used in packaging*

Quality control in the production of flat pack furniture

Flat pack furniture is manufactured from a sheet material such as chipboard which can be adversely affected by damp or humid conditions. If the chipboard is damaged in any way, the final product will be below the standards expected by the company. Prior to manufacture, the chipboard must be checked, and whilst in storage must be kept in a stable environment. During the manufacturing process, regular QC checks will need to be made to ensure the final quality of the product meets the expected standard.

THINK ABOUT THIS!

What might the consequences be of manufacturing flat pack furniture without applying QC procedures?

Total quality management

Total quality management (TQM), often referred to as total quality control (TQC), is the strategic integrated system for achieving customer satisfaction by applying QA procedures at every stage of the production process. TQM is based on all members of an organisation participating in the continual improvement of processes, products, services and the overall culture in which they work. Each department in a company is treated as a client, therefore ensuring high standards of service and attention to detail when dealing between departments. For example a production team must produce a high-quality component that the assembly team know is quality assured and will therefore fit perfectly.

The British Standards Institute (BSI) operates a quality management system to which any organisation can be accredited. This system, known as the ISO 9000 series of standards, recognises companies who can demonstrate high standards of consistency in all of their procedures. The ISO 9000 series is the world's most established quality framework, and is used by over 250,000 organisations in 161 countries. If an organisation is accredited with this standard, the customer is assured of the quality of the product and service.

BSI™

Figure 2.29 *BSI logo*

Sector	Benefits
Customers and users	• Receive products that conform to the requirements • Are dependable and reliable • Are available when needed • Easily maintained
People in the organisation	• Better working conditions • Increased job satisfaction • Improved health and safety • Improved morale
Owners and investors	• Increased return on investment • Improved operational results • Increased market share • Increased profits
Society	• Fulfilment of legal and regulatory requirements • Improved health and safety • Reduced environmental impact • Increased security

Table 2.23 *Benefits of the BS EN ISO 9000 series of standards*

Problem	Description	Quality control
Veneer bubbles	When veneer is put onto sheet material such as chipboard, there is a danger that the veneer will not stick and this causes bubbles to occur between the chipboard and the veneer.	Check and make sure that glue covers the whole surface of the sheet material and that when placed in a vacuum pack the vacuum is complete.
Cut edges not straight	It is vital that for the furniture to fit together all edges need to be straight. If this is not done, edges will not butt up to each other and the result will be a poor finish.	Periodically check that all guides and fences on machines are set to the correct sizes and distances.
Holes not positioned	If holes are not positioned correctly on flat pack furniture then nothing will fit together properly.	Regular checking of any jigs is vital for the accurate assembly of the product.

Table 2.22 *Possible quality control checks used during the manufacture of flat pack furniture*

Quality standards

As a mark of quality, manufacturers often seek validation of their products by applying for formal 'quality standards'. There are three main types:

- **National Standards**, for example British Standards (suffix BS), are produced by a country's national standards body (NSB). In the UK, British Standards (BS) are developed together with the UK Government, businesses and society. Some are enforced by regulation, but most standards are voluntary.
- **European Standards** (suffix EN) are produced by the European Committee for Standardisation (CEN), whose members are the NSBs of the European Union countries. In the UK, they are adopted as British Standards.
- **International Standards** (suffix ISO) are produced by the International Organisation for Standardisation (ISO), whose members are the NSBs of countries all over the world. BSI British Standards is a leading member of the ISO and represents the UK's interest in the development of international standards. BSI also decides which international standards to adopt as British Standards (BS ISO).

Kitemark and CE marking

The Kitemark® scheme has been in place since 1902 and over the years has become an important symbol of quality and covers a wide variety of products from electrical components to double glazing. Having a Kitemark® associated with a product or service certifies that it complies with particular standards. It means that any product or service that displays a Kitemark® has been

Figure 2.30 *European CE marking for quality*

through a number of rigorous quality processes and that consumers can buy products or services knowing that they have reached required standards.

The letters 'CE' on a product are the manufacturer's claim that the product meets the requirements of all relevant European Directives. Many products are covered by these directives and for some, such as toys, it is a legal requirement to have a CE mark. This shows that the product achieves a minimum level of quality, and ensures it can be moved freely throughout the European Single Market.

www.bsieducation.org
www.bsi-global.com

Health and safety

Getting started

Health and safety is a very important subject. Employers are legally required to minimise the risks to their employees and in turn employees need to take reasonable care when carrying out their jobs. The school design and technology department is no different. What would happen if there were no signs on machinery or you were simply allowed to 'do what you want' in a workshop? Acting in an irresponsible way that might cause an injury or illness to yourself or someone else is a criminal offence that might lead to prosecution.

Health and Safety at Work Act (1974)

Under this Act of Parliament, employers are legally required to do all that is reasonably practicable to ensure the health, safety and welfare at work of employees, and the health and safety of non-employees such as students and visitors to a school. The following regulations are procedures to safeguard the risk of injury to people.

Personal and protective equipment (PPE)

Personal Protective Equipment at Work regulations (1992) state that employers have basic duties concerning the provision and use of personal protective equipment (PPE) at work. PPE is defined in the regulations as 'all equipment (including clothing providing protection against the weather) which is intended to be worn or held by a person at work and which protects him against one or more risks to his health or safety'. These can include safety helmets, gloves, eye protection, high-visibility clothing, safety footwear and face masks or respirators.

The main requirement of the regulations is that PPE is to be supplied and used at work wherever there are risks to health and safety that cannot be adequately controlled in other ways. The regulations also require that PPE is:

- properly assessed before use to ensure it is suitable
- maintained and stored properly
- provided with instructions on how to use it safely
- used correctly by employees.

Risk	Hazards	Personal protective equipment (PPE)
Eyes	Chemical or metal splash, dust, projectiles, gas and vapour, radiation	Safety spectacles, goggles, face shields, visors
Head	Impact from falling or flying objects, risk of head bumping, hair entanglement	A range of helmets and bump caps
Breathing	Dust, vapour, gas, oxygen-deficient atmospheres	Disposable filtering face-piece or respirator, half- or full-face respirators, air-fed helmets, breathing apparatus
Protecting the body	Temperature extremes, adverse weather, chemical or metal splash, spray from pressure leaks or spray guns, impact or penetration, contaminated dust, excessive wear or entanglement of own clothing	Conventional or disposable overalls, boiler suits, specialist protective clothing, e.g. chain-mail aprons, high-visibility clothing
Hands and arms	Abrasion, temperature extremes, cuts and punctures, impact, chemicals, electric shock, skin infection, disease or contamination	Gloves, gauntlets, mitts, wrist-cuffs, armlets
Feet and legs	Wet, electrostatic build-up, slipping, cuts and punctures, falling objects, metal and chemical splash, abrasion	Safety boots and shoes with protective toe caps and penetration-resistant mid-sole, gaiters, leggings, spats

Table 2.24 *Hazards and types of personal protective equipment (PPE)*

Signage

The Safety Signs (Signs and Signals) regulations (1996) require employers to display an appropriate safety sign and instruction wherever a significant risk or harm cannot be avoided or reduced by other means. These regulations bring into force a European Directive whose purpose is to encourage the standardisation of safety signs throughout Europe so that safety signs, wherever they are seen, have the same meaning.

The regulations cover various means of communicating health and safety information. These include the use of illuminated signs, acoustic signals such as fire alarms, and traditional signboards such as prohibition, warning and fire safety signs, for example signs for fire exits and fire-fighting equipment.

Health and safety signage		Example
	Prohibition signs are used to prohibit actions to prevent personal injury and the risk of fire.	
	Mandatory signs convey action that must be taken, e.g. procedures in case of fire.	
	Warning signs are to warn personnel of possible dangers in the workplace.	
	Safe condition signs show directions to areas of safety and medical assistance and indicate where a safe area, safety equipment or first aid equipment is located.	
	Fire equipment signs show the location of fire equipment and compliance with Fire Precautions.	

Table 2.25 *Standard health and safety signage*

Warning symbols

Warning symbols are placed on products to provide health and safety information for the consumer. An example is British Standards (BS) EN 71 which is concerned with the safety of toys, of which *Part 6: Graphical Symbol for Age Warning Labelling*, covers age warning symbol labelling and specifies the requirements of the symbols used on toys not suitable for children under the age of three.

Many warning symbols appear on the packaging of adhesives and domestic cleaning products along with additional safety instructions that outline any potential risks to users.

Not suitable for children under 3 years (36 months)

Irritant to eyes and skin

Figure 2.31 *Warning symbols on packaging*

WEBLINKS:

www.bsieducation.org – British Standards Institute (BSI)
www.btha.co.uk – British Toy and Hobby Association

THINK ABOUT THIS!

Why is it important for the packaging of toys to carry an age warning symbol? What are the risks to a young child's health and safety by not paying attention to such warnings?

Health and Safety Executive risk assessments

The Health and Safety Executive (HSE) lays down government guidelines for health and safety issues within the workplace, including schools. The HSE states that all places of work must carry out risk assessments of their facilities to identify any potential hazards to employees or students, and to put in place control measures to reduce the risk of injury.

The HSE outlines its Five Steps to Risk Assessment:

1. Identify the hazard.
2. Identify the people at risk.
3. Evaluate the risks.
4. Decide upon suitable control measures.
5. Record risk assessment.

www.hse.gov.uk

The Health and Safety Commission is responsible for health and safety regulation in the UK.

Using computers

Computers are often used as a design tool. This can involve large amounts of time sitting at a computer workstation, looking at a monitor or screen, typing and using a mouse. These are all potential hazards. The HSE, together with the European Union's 'VDU Directive', has regulations and guidance on working at a computer workstation.

Repetitive strain injury

Repetitive strain injury (RSI) is a medical condition affecting muscles, tendons and nerves in the arms and upper back. It occurs when muscles in these areas are kept tense for very long periods of time, due to poor posture and/or repetitive motions.

Hazard	Risk	Control measure
Potential (of risk) from a substance, machine or operation.	Reality (of harm from the hazard).	Action taken to minimise the risks to people.

Figure 2.32 *What is the difference between a hazard and a risk?*

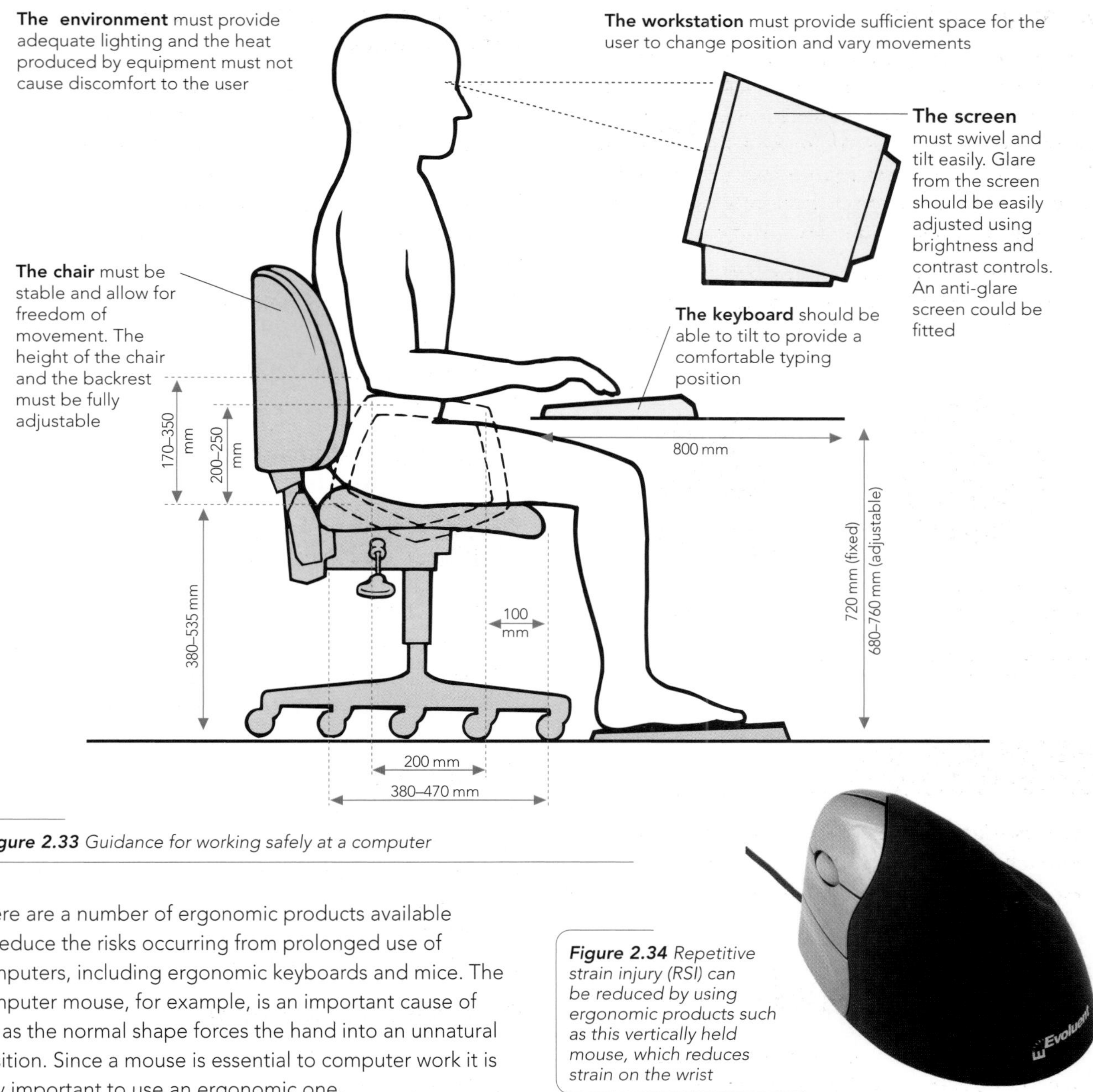

The environment must provide adequate lighting and the heat produced by equipment must not cause discomfort to the user
The workstation must provide sufficient space for the user to change position and vary movements
The screen must swivel and tilt easily. Glare from the screen should be easily adjusted using brightness and contrast controls. An anti-glare screen could be fitted
The chair must be stable and allow for freedom of movement. The height of the chair and the backrest must be fully adjustable
The keyboard should be able to tilt to provide a comfortable typing position
170–350 mm
200–250 mm
800 mm
380–535 mm
100 mm
720 mm (fixed)
680–760 mm (adjustable)
200 mm
380–470 mm

Figure 2.33 *Guidance for working safely at a computer*

There are a number of ergonomic products available to reduce the risks occurring from prolonged use of computers, including ergonomic keyboards and mice. The computer mouse, for example, is an important cause of RSI as the normal shape forces the hand into an unnatural position. Since a mouse is essential to computer work it is very important to use an ergonomic one.

Figure 2.34 *Repetitive strain injury (RSI) can be reduced by using ergonomic products such as this vertically held mouse, which reduces strain on the wrist*

Hazard	Risk	People at risk	Control measure
Using a computer	Repetitive strain injury (RSI)	User	• Keyboard should tilt to provide a comfortable typing position • Use an ergonomic keyboard with wrist support • Use an ergonomic mouse • Take regular breaks to rest hands
	Eye strain	User	• Adjust glare from monitor using brightness and contrast controls • Use of an anti-glare screen fitted to monitor to reduce screen flicker • Tilt or swivel monitor to reduce reflections • Take regular breaks to rest eyes

Table 2.26 *Part of a risk assessment for using a computer*

Hazard	Risk	People at risk	Control measure
Using a pillar drill	Damage to eyes from flying debris	User/people in immediate area	• Use appropriate PPE, for example safety spectacles or goggles • User fully briefed on use of machine (general machine safety), i.e. guards in position • Appropriate supervision by teacher or technician • Other students should wait behind marked yellow lines or barriers when not using machine
	Cuts from metal shavings	User	• Work clamped securely in vice (never held in hand) to prevent work from catching and spinning • Use of appropriate PPE, for example gloves • Use a small 'stick' to remove large spiral shavings • Place shavings in appropriate disposal container

Table 2.27 *Part of a risk assessment for using a pillar drill*

Workshop practices

When undertaking practical work in the workshop it is likely that you will be given the opportunity to work on a range of tools and machines. Obviously, there will be a wide range of potential hazards when using machinery, power tools and equipment. The school, or college, will have carried out a detailed risk assessment for each piece of equipment, which should be clearly displayed for your information.

THINK ABOUT THIS!

In small groups, carry out a series of risk assessments for different workshop machinery, equipment and processes. Collate your work to build up a comprehensive set of risk assessments that you can use in your coursework projects.

Control of Substances Hazardous to Health regulations

The Control of Substances Hazardous to Health (COSHH) regulations place a duty on employers to make an assessment of risks for work involving exposure to substances hazardous to health. Steps must be taken to prevent or control adequately the exposure of employees and others to these substances. Hazardous substances include:

- substances used directly in work activities such as adhesives, paints and cleaning agents
- substances generated during work activities such as fumes from soldering and welding
- naturally occurring substances such as dust
- biological agents such as bacteria and other micro-organisms.

Step		Action
1	Assess the risks	Assess the risks to health from hazardous substances used in or created by workplace activities.
2	Decide what precautions are needed	Do not carry out work that could expose employees to hazardous substances without first considering the risks and the necessary precautions, and what else is needed to comply with COSHH.
3	Prevent or adequately control exposure	Prevent employees being exposed to hazardous substances. Where preventing exposure is not reasonably practicable, then it must be adequately controlled.
4	Ensure that control measures are used and maintained	Ensure that control measures are used and maintained properly and that safety procedures are followed.
5	Monitor exposure	Monitor the exposure of employees to hazardous substances, if necessary.

Table 2.28 *Health and Safety Executive (HSE) guidance on COSHH regulations*

6	Carry out appropriate health surveillance	Carry out appropriate health surveillance where assessment has shown it necessary or where COSHH sets specific requirements.
7	Prepare plans and procedures to deal with accidents, incidents and emergencies	Prepare plans and procedures to deal with accidents, incidents and emergencies involving hazardous substances, where necessary.
8	Ensure employees are properly informed, trained and supervised	Provide employees with suitable and sufficient information, instruction and training.

Many adhesives are solvent-based, containing volatile organic compounds (VOCs) that give off vapours that can cause dizziness and nausea. Because of this, these substances are extremely hazardous to use within confined indoor areas such as workshops or classrooms. It is important that thorough risk assessments are carried out and the appropriate action taken to minimise the risks.

www.coshh-essentials.org.uk – COSHH Regulations

Hazard	Risk	People at risk	Control measure
Use of solvent-based adhesives	Burns from corrosive adhesives	User	• Use appropriate PPE, including gloves and eye protection • Users fully briefed on safe use of adhesives • Appropriate supervision by teacher or technician • Wash area immediately with warm soapy water and seek medical attention • Eyes: seek medical attention immediately; use an eye bath to wash eyes
	Inhalation of VOC vapours	User/people in immediate area	• Use only in well-ventilated areas, i.e. use extraction or open external windows/doors • Appropriate supervision by teacher or technician • Use of face-mask or respirator • If dizziness and nausea occur vacate area immediately and seek medical attention
	Storage	Technicians and teaching staff	• Store in a secure metal cupboard • Cupboard easily identifiable (yellow) with appropriate safety signage clearly displayed • Staff to be fully briefed as to safe storage of adhesives • Checks by technician on a regular basis

Table 2.29 *Part of a risk assessment for storage and use of solvent-based adhesives*

Exam**Café**

Teacher area

Revision summary

It is important to note that the questions asked by the examiner in this exam paper will cover aspects from **all four** sections of this unit. No paper will ever focus upon one section entirely. Therefore, it is vital that you have a secure knowledge and understanding across all four sections.

You should give yourself plenty of opportunities to answer examination-style questions throughout the course so you are prepared for the final examination. Use the sample assessment materials (SAMs) and past exam papers provided by Edexcel and the questions in this handbook.

Don't forget – if in doubt, ASK! Your teacher is there to help you understand the theory in this unit. Have a good, long think about appropriate questions to ask your teacher – it might be a good idea to discuss a problem with your peers first to see if they can explain it more clearly.

Finally, keep a set of well-ordered and legible revision notes, which will help you to learn key topics and provide you with something to refer back to when in doubt.

Refresh your memory

Revision checklist

- Make sure that you have answered all the questions at the end of this section.
- Make sure that your revision notes are well ordered, clear and up to date.
- Use the web links to read around each key topic so that you are well informed.
- Use SAMs and past papers to practise your exam technique.
- Discuss any problems with your peers or teacher – don't keep them to yourself!

Get the result!

Tips for answering questions

Always read each question carefully before you respond. It might be a good idea to use a piece of scrap paper to outline your response if you think you have enough time.

Always look at the amount of marks awarded for each question in brackets. This will give you a good indication of how many points need to be raised in your response. As a general rule of thumb, look at the following command words and what you have to do in order to gain the marks:

Give, state, name	(1 mark)	These types of questions will usually appear at the beginning of the paper, or question part, and are designed to ease you into the question with a simple statement or short phrase.
Describe, outline	(2+ marks)	These types of questions are quite straightforward. They ask you to simply describe something in detail. Some questions may also ask you to use notes and sketches; you can gain marks with the use of a clearly labelled sketch.
Explain, justify	(2+ marks)	These types of questions will be commonplace in this exam. They are asking you to respond in detail to the question – no short phrases will be acceptable here. Instead, you will have to make a valid point and justify it.
Evaluate	(4+ marks)	These types of questions will appear towards the end of the paper and are designed to stretch and challenge the more able student. They are awarded the most marks because they require you to make a well-balanced argument, usually involving both advantages and disadvantages.

Ask the examiner: Worked examples

The following four questions should demonstrate the style of questions using the different types of command words. The places where marks have been awarded are indicated in brackets. These are referred to as 'trigger points' and are parts of the examiner's mark scheme where it is expected marks will be awarded.

Exam question 1

Give **two** reasons why a park bench would be manufactured from a hardwood.

(2 marks)

This question is a straightforward 'give' question, so short statements are acceptable and they do not have to be justified.

Joseph

1. *Hardwood is durable.* **(1 mark)**
2. .... **(1 mark)**

Here, the student has given one relevant reason for using hardwood. However, he has not even attempted a second response. It is really important that you attempt all questions – even if you have to make an educated guess!

Elizabeth

1. *Hardwoods resist weathering more than softwoods.* **(1 mark)**
2. *Hardwoods tend to be stronger than softwoods.* **(1 mark)**

(2 marks)

Two well-constructed sentences that give two appropriate reasons why hardwoods are used for park benches earn full marks.

Exam question 2

Explain **two** reasons why polystyrene (PS) is used for the casing of many electrical products, such as mobile phones.

(4 marks)

This question asks you to apply your knowledge and understanding of polymers (polystyrene, to be specific) to a familiar product. In this question you are asked for two explanations worth two marks each. In other words, two fully justified points.

Louise

1. *You can get polystyrene in a wide range of colours* **(1 mark)** *so your mobile phone can be different colours.*
2. *Polystyrene is lightweight.* **(1 mark)**

(2 marks)

Here, the student has given two acceptable properties of polystyrene that make it suitable for use in a mobile phone. However, the first answer is not fully justified and the second offers no justification at all.

Sally

1. *Polystyrene is a thermoplastic,* **(1 mark)** *which means that it can be easily formed* **(1 mark)** *into a casing using injection moulding.*
2. *Polystyrene is lightweight,* **(1 mark)** *so it is ideal for portable devices* **(1 mark)** *such as mobile phones.*

(4 marks)

This student achieves full marks for this question as both properties of polystyrene are fully justified when applied to its use in mobile phones.

Exam question 3

Justify the use of plywood rather than a piece of solid timber when considering the manufacture of the top for a dining room table.

(4 marks)

This question relates to the laminates section of the specification. You must apply your subject knowledge and understanding of plywood and its application as a suitable sheet material. You should then be able to compare and contrast it with the second material, solid wood. You should make two fully justified points. The marks in brackets indicate the number of marking points the examiner is looking for.

Jonathan

Plywood comes in large sheets **(1 mark)** *and it is safer.* **(1 mark)**

The student has given one relevant point, plywood does come in large sheets and is therefore suitable for tabletops. The second point is not relevant. It may be safer, but the point is not relevant to the question and no justification has been offered.

Tim
Plywood comes in thin sheets **(1 mark)** *that can easily be cut by hand* **(1 mark)**. *It is a very stable material* **(1 mark)** *because it is made up of layers of veneers arranged with the grain at 90°* **(1 mark)**.

(4 marks)

This response gets full marks because each point is relevant and fully justified.

Exam question 4

Evaluate the use of mass production to manufacture many consumer goods.

(6 marks)

This question asks for an 'evaluation' of a topic so the response must look at both sides of the argument – using both advantages and disadvantages. It is a type of question that is supposed to stretch students because it is more open ended.

Annie
Mass production is a good way of producing thousands of products because people want to buy them so they have to cater for the demand **(1 mark)**. *When thousands are produced the cost of manufacture reduces so products are cheaper* **(1 mark)**. *Products aren't as dear so more people will buy them and that's why mass production is good.*

(2 marks)

Here, the student has started well with two relevant points (although not really justified enough to gain two marks each) but soon runs out of steam. The last sentence repeats an earlier point – 'products are cheaper' is the same as 'products aren't as dear'.

Ruthie
Mass production is a suitable scale of production for many consumer goods because it takes advantage of economies of scale **(1 mark)** *and buying materials and components in bulk makes them cheaper* **(1 mark).** *Mass production uses automated machinery on production lines,* **(1 mark)** *which is a highly efficient* **(1 mark)** *way of manufacturing on a large scale.*

However, mass production can have many negative effects upon the workforce. Many workers are low skilled **(1 mark)** *and simply mind machines. This can lead to poor job satisfaction* **(1 mark)** *due to the mundane and repetitive nature of the job.*

(6 marks)

This response gains full marks as it addresses both the advantages and disadvantages of mass production. The response is succinct and well structured, containing six trigger points from the examiner's mark scheme.

Practice questions

1. Materials are selected for products for various reasons:
   (i) Name a suitable material for the metal frame of a lightweight garden chair. **(1 mark)**
   (ii) Name a suitable wood for a garden bench. **(1 mark)**
2. Give **one** advantage and **one** disadvantage of using mild steel for a bracket to hold a hanging flower basket. **(2 marks)**
3. Medium density fibreboard (MDF) is extensively used in school workshops. Explain **one** advantage and **one** disadvantage of using medium density fibreboard (MDF) in the workshop. **(4 marks)**
4. Explain **two** reasons why polyethylene terephthalate (PET) is used in the manufacture of a fizzy drinks bottle. **(4 marks)**
5. Outline the following stages in the commercial production of the case of a hand held electric drill.
   (i) Preparation **(2 marks)**
   (ii) Processing/production **(4 marks)**
   (iii) Assembly **(2 marks)**
   (iv) Finishing **(2 marks)**
6. Plastic bottles are produced using the blow moulding process. Using notes and sketches explain the blow moulding process. **(6 marks)**
7. Justify the use of quality control (QC) in the manufacturing process. **(4 marks)**
8. Explain how the risks could be reduced when using a metalworking centre lathe. **(4 marks)**
9. Evaluate the use of one-off production to produce products. **(6 marks)**
10. Evaluate the advantages and disadvantages of thermoplastics and thermosetting plastics. **(6 marks)**

Unit 3:

Designing for the Future

Summary of expectations

1 What to expect

In this unit, you will develop your knowledge and understanding of a range of modern design and manufacturing practices and contemporary design issues. The modern designer must have a good working knowledge of the use of information and communication technology (ICT) and systems, and control technology in the design and manufacture of products. They must also be aware of the important contributions of designers from the past, which may provide inspiration for future design.

It is increasingly important that you develop an awareness of the impact of design and technological activities upon the environment. Sustainable product design is a key feature of modern design practices.

2 How will it be assessed?

Your knowledge and understanding of topics in this unit will be externally assessed through a 2-hour examination paper set and marked by Edexcel. The exam paper will be in the form of a question and answer booklet consisting of short-answer and extended-writing type questions.

The total number of marks for the paper is 70.

3 What will be assessed?

This unit is divided into four main sections, with each section outlining the specific knowledge and understanding that you need to learn:

3.1 Industrial and commercial practice

- Information and communication technology
- Biotechnology

3.2 Systems and control

- Manufacturing systems
- Computer-integrated manufacture
- Robotics and artificial intelligence
- Flow charts

3.3 Design in context

- The effects of technological changes on society
- Influences of design history on the development of products
- Form and function
- Anthropometrics and ergonomics

3.4 Sustainability

- Life-cycle assessment
- Cleaner design and technology
- Minimising waste production
- Renewable and non-renewable sources of energy
- Responsibilities of developed countries

4 How to be successful in this unit

To be successful in this unit you will need to:

- have a clear understanding of the topics covered in this unit
- apply your knowledge and understanding to a given situation or context
- use specialist technical terminology where appropriate
- write clear and well-structured answers to the exam questions that target the amount of marks available.

5 How much is it worth?

This unit is worth 40 per cent of the full Advanced GCE.

Unit 3	Weighting
A2 level (full CGE)	40%

Industrial and commercial practice

Getting started

This section follows on from *Unit 2: Materials and components* (*see* pages 25-40) and develops your knowledge and understanding of modern design practice. We all know that the modern world relies upon computers and that the modern workforce must be computer literate – but what are the advantages and disadvantages of using these systems? Biotechnology is a controversial new technology where living organisms are genetically modified to produce products for a specific use – is this right? What impact do these modern technologies have upon our lives?

Information and communication technology

Information and communication technologies (ICT) are increasingly being used by manufacturers when designing, manufacturing and selling products. With modern technologies, companies are able to take all aspects of the designing, selling and manufacturing processes to an unprecedented scale allowing every aspect of the process to be undertaken worldwide.

Electronic communications

E-mail

Information and communication technology (ICT) has improved the reach (level of communication across a network) and range (types of data transfer available) of electronic communications. E-mail is the simplest form of electronic communication and has a comparatively low level of reach and range when it is used for messaging or transferring documents. However, it has proved invaluable in rapid communications between designers, manufacturers, retailers and consumers due to its ease of use and widespread access through Internet connections. For these reasons it has all but replaced the postal system. There are issues of security and privacy when using e-mail and limitations on the size of attachments, which often restrict its use, but to the majority of people it is now their preferred way of communicating.

Electronic data interchange

Electronic data interchange (EDI) is a new way for companies to do 'paperless' business using a process that transfers business documents through a computer network, rather than via the postal system. Many modern companies use EDI as a fast, inexpensive and secure system of sending purchase orders, invoices, design and manufacturing data, etc. For example, some manufacturers use EDI to transmit large, complex computer-aided design (CAD) drawings and multinational companies use EDI to communicate between locations worldwide. EDI can also be used to transmit financial information and payment in electronic form. However, the transfer of files requires that the sender and receiver agree upon a standard document format for the document that is to be transmitted.

Advantages	Disadvantages
• Quick, easy and convenient means of communicating around the world • Widespread usage (anyone with a computer connected to the Internet) • E-mail exchanges can be saved as a dated record of correspondence • Documents can be attached electronically and can be saved and edited easily	• Impersonal and some messages can be misinterpreted • Influx of messages to inbox increases time to read and respond • 'Spamming' of unsolicited commercial e-mails, often with inappropriate content • Privacy and security issues as messages can be intercepted and read • Limitations on size of attachments • System can crash and all files lost

Table 3.1 *Advantages and disadvantages of e-mail*

The EDI process starts with a trading agreement between a company and their trading partner. Joint decisions have to be made regarding the standard to be used, the information to be exchanged, how the information is to be sent, and when information will be sent. To send a document, EDI translation software is used to convert the document format into the agreed standard. The translator creates and wraps the document in an electronic envelope with an identification code and sends it to the partner's mailbox. The document is retrieved from their mailbox and an EDI translator opens the envelope and translates the data from the standard form to their application's format. The translator ensures that the data sent by one company is converted into a format that another can use.

Advantages	Disadvantages
• Saves money by eliminating the need for processing paper documents. • Saves time as information is transferred digitally. • Improves customer service as business documents are transferred quickly with fewer errors. • Expands customer base due to improved customer service through efficient EDI processes.	• Incompatibility with some companies due to range of standard document formats. • Standards updated regularly, which causes problems with different versions in use. • Expensive to initially set up. • Limits trading to only companies with EDI.

Table 3.2 *Advantages and disadvantages of EDI for business*

Integrated Services Digital Networks and broadband

The development of Integrated Services Digital Networks (ISDNs) and, more recently, broadband technology, means that huge amounts of information can be transferred across computer networks at far greater speeds than ever before. The purpose of ISDN is to provide fully integrated digital services to users comprising digital telephony and data-transport services through existing telephone networks. ISDN involves the digitisation of the telephone network, which enables voice, data, text, graphics, music and video to be transmitted at high speeds over existing telephone lines. The emergence of ISDN represented an effort to combine subscribed services such as Internet service provision and telecommunications into one package. It enabled users to have additional phone lines installed so that the Internet could be used at the same time as the telephone without callers receiving a 'busy' signal.

Local Area Networks

The Internet is a global network and as such can be accessed anywhere in the world. Local Area Networks (LANs) are as the name suggests networks that operate over a much smaller area. This area might be as confined as a single office, or building, or at the most a number of offices situated perhaps a few kilometres apart. The majority of LANs connect workstations or personal computers (PCs). Each workstation or PC has a central processing unit (CPU) and is able to undertake tasks and operate programs independently but it is also able to access data from anywhere on that particular LAN.

In an office, for example, workers can log onto a machine and undertake work and then communicate that work to any other workstation on that particular LAN. Not only can workers share information on the LAN but they can also share devices such as scanners or printers. They are able to send e-mails and have instant 'chat' facilities with anyone on the same LAN. Because of the small nature of the network, LANs tend to be faster than using the Internet because data can be transferred at a much higher rate. A typical data transfer rate on a LAN might be up to 100 megabits per second. Very often wireless and fibre optic technologies are used. Added to all these features is the fact that because the system is 'local', security is easier to maintain on a LAN system.

FACTFILE:

Key points of Local Area Networks
• Local (i.e. one building or group of buildings) • Controlled by one administrative authority • Assumes other users of the LAN are trusted • Usually high speed and is always shared

Advantages	Disadvantages
Speed • Sharing and transferring files within networks are very rapid **Cost** • Individually licensed copies of many popular software programs can be costly, while networkable versions are available at considerable savings **Security** • Sensitive files and programs on a network are password protected **Centralised software management** • Software can be loaded on one computer (the file server), eliminating the need to spend time and energy installing updates and tracking files on independent computers throughout the building **Resource sharing** • Resources such as printers, fax machines and modems can be shared **Electronic mail** • Electronic mail on a LAN can enable staff to communicate within the building without having to leave their desks **Flexible access** • Access their files from computers throughout the firm **Workgroup computing** • Workgroup software (such as Microsoft BackOffice) allows many users to work on a document or project concurrently	**Faults** • Server faults stop applications being available • Network faults can cause loss of data • Network faults could lead to loss of resources **Users** • User work dependent upon network **Misuse** • System open to hackers **Resources** • Scanners and printers could be located too far from users

Table 3.3 *Advantages and disadvantages of LANs*

Global networks (Internet)

The Internet has revolutionised the way in which information is passed around the globe. With the continuing development of the web, the transmission of digital information has become possible on an unprecedented scale. Where LAN systems are limited to a particular company or location, the Internet has opened up worldwide possibilities. It means that a company is able to keep in touch with its branches and associates anywhere in the world literally at the touch of a button. It means that information can be passed around instantly, opening up the possibilities of things like design being undertaken in Europe and manufacturing in Asia.

However, there are downsides. One of the great benefits of a LAN system is that it can be kept almost totally secure and sensitive information kept within the company. The links that a LAN system has with the outside world are controllable and highly monitored. With the Internet the security is weaker and there is always a danger that hackers could get into what appear to be secure installations and gain confidential information.

Advantages	Disadvantages
Communication • Fast and reliable communications • Advent of the truly global village **Information** • Treasure trove of information • Search engines assist in finding information • Every kind of technical support available • Every kind of trade or law or market information available **Services** • Banking • Reservations **E-Commerce** • Online banking • Online shopping	**Theft of personal information** • Danger of identity theft • Access to credit cards and bank accounts **Spamming** • Receiving unsolicited e-mails and information **Virus threat** • Virus is a program which disrupts the normal functioning of computer systems

Table 3.4 *Advantages and disadvantage of global networks*

Videoconferencing

Integrated Services Digital Network (ISDN) and broadband have enabled high-speed data and communications transfer, which can be used in a videoconferencing system. A videoconference allows two or more locations to interact using two-way video and audio transmissions simultaneously, enhancing communications and speeding up the decision-making process by eliminating the need for time-consuming travel to meetings, which might be across the other side of the world.

A videoconferencing system includes a video camera to capture images, a screen to view images, microphones to pick up sound and loudspeakers to play sounds. Data is transferred via the Internet using ISDN or broadband. There are two types of videoconferencing systems:

- **Dedicated systems** have all the required components packaged into a single console, including a high-quality remote-controlled video camera.
- **Desktop systems** are add-ons to normal personal computers such as webcams and microphones, transforming them into videoconferencing devices.

Multipoint videoconferencing allows for simultaneous videoconferencing among three or more remote points using a multipoint control unit (MCU) as a bridge that interconnects calls from several sources. This enables three or more people to sit in a 'virtual' conference room and communicate as if they were sitting next to each other.

Figure 3.1 *Multipoint videoconferencing allows people in three or more locations to communicate with each other simultaneously*

Advantages	Disadvantages
• Eliminates the need for travel to other countries, saving both time and money. • Visual information can be communicated as part of the conversation. • Accelerates the decision-making process as presentations can be made to several people simultaneously. • Remote diagnostics available with technicians in other countries able to solve problems. • Corporate training of many staff at the same time.	• Synchronisation of time of meeting in different times zones across the world. • Connection can often fail. • Lack of eye contact with others in meeting can hinder conversation or intent. • 'Camera shyness' can hinder presentations due to pressures of being filmed and often recorded.

Table 3.5 *Advantages and disadvantages of videoconferencing*

Electronic information handling

Market analysis

Market analysis is an investigation of a market that is used by a company when planning future activities. A market analysis is used by companies to help make a wide range of decisions. For example, it could be used to look at the available market for a product or it may be used to predict the expansion or contraction of the workforce. It can influence issues ranging from how a new product is advertised through to estimates on capital outlay.

You will be familiar with some of the ways in which market analysis is undertaken from your design and make projects that you have produced. In those you are expected to undertake research. This research is then used to produce a product specification.

The raw data for market analysis can be collected in a number of ways. The most obvious way is to use a questionnaire. Once the data has been collected, specialist software processes it. Once the data has been processed, it can be sorted so that companies can interpret the results. On the basis of the results, decisions are taken about the nature of the marketing plan and about how best to satisfy the needs of the intended customers. Market analysis is essential in order to ensure that new products attain a competitive edge over competitors.

Specification development

When developing a specification for a new product, a whole range of factors need to be taken into account. For example, issues such as form, function, user

Design specification	Sets the parameters of what is to be designed. Much of what it contains will be the result of market research. Computer-Aided Market Analysis (CAMA) systems can be interrogated in a number of ways (qualitative, quantitative, trend etc.). The results of these interrogations will provide a focus area for a design specification, i.e. highlight features and aspects that are most important to the consumer and therefore a guide to the content of the specification.
Manufacturing specification	This could be the working, or scale drawings, of a product that contains all the information required to manufacture the product. A designer can generate these on a CAD system which has many advantages over traditional drawing methods. Computer-Aided Specification of Products (CASP) draws on artificial intelligence expert systems to suggest modifications and alterations to the design based on its knowledge of materials, virtual testing, manufacturing facilities available and ease of assembly. This results in a manufacturing specification that is both generated and influenced by computer systems rather than being just the result of a designer's decisions.

Table 3.6 *Types of specifications that can be generated using computer technology*

requirements, performance requirements, materials and processes, the scale of production and costs all need to be considered. These are all factors which you are familiar with as they must all be thought through when producing any design and make project.

For a school project, the development of the specification is relatively easy to undertake. In an industrial situation, the issues involved will be much more complex and therefore computer programs are used to help produce product specifications. The process involves defining all the parameters which relate to the product. Using information such as anthropometric data, material limitation and even market research, the computer establishes factors such as components required for a particular product, materials, measurements and even building and construction data. The advantage of such a system is that the process is swift, thus reducing the lead time for a product. It also means that, as the information is held on computer, all those who have need of the most up-to-date specification can gain easy access.

There are two situations where specifications are used in the design process: design specification and manufacturing specification, both of which can be generated with the aid of computers.

LINKS TO:

Unit 3.2 Systems and control: Computer-integrated manufacture.

Automated stock control

'Just in time'

'Just in time' (JIT) manufacturing is a system used by companies to reduce costs. The philosophy of JIT manufacturing is related to efficiency. A product is only manufactured when that product is actually required. In traditional manufacturing components are taken to the next stage of production as soon as they are ready. In JIT production each stage must be provided with the number of components required. By only producing items when they are actually needed the quality and efficiency of the manufacturing process are improved. It can also lead to higher returns for the company itself.

The idea is that a re-order level is set within stock control and new stock is only ordered when that level is reached. There is therefore no overordering, which saves space and again increases efficiency. An example of JIT can be seen in the manufacture of cars. Manufacturers only buy in materials for the immediate use of the manufacturing process. As a consequence the process of manufacturing is smooth, with just the right amount of materials being delivered to the factory at the correct time. The turnover is rapid and the amount of money tied up in raw materials and components is reduced. Vehicles are usually built to order, and that reduces the problems of producing cars that can never be sold, thus reducing the risk to the company even further.

Production scheduling and production logistics

When undertaking the manufacture of any product it is important that all the resources and the sequencing of tasks are allocated in the correct way to ensure efficient production. This process can be very complex.

Computer-based production scheduling and production logistics help make the production process smooth. The idea is that if there are small issues or hiccups in the process the computer can iron them out. By producing a production schedule a company can determine whether or not a delivery promise can be met. It gives the workers undertaking the manufacturing process itself a statement of what should be done so that their productivity might be measured, therefore maximising the potential of the workforce.

Advantages of computer-based scheduling and production
• Flexible and easily adaptable if product mix or quantity changes • Minimise work in progress and reduce inventory • Maintain balance on production line • Raise productivity levels

Table 3.7 *Advantages of computer-based scheduling and production*

Flexible manufacturing systems

Flexible manufacturing systems (FMS) are manufacturing systems that have been set up that allow a company some degree of flexibility to enable it to react to, and then instigate, changes fairly rapidly to the manufacturing process. These changes might be either predicted or unpredicted. This flexibility falls into two main categories, machine flexibility and routing flexibility.

Flexibility	
Machine flexibility	• System's ability to change to produce new product types • Ability to change order of operations
Routing flexibility	• Ability to use multiple machines • Ability to absorb changes such as capacity and capability

Table 3.8 *The two types of flexibility in manufacturing*

Such flexible manufacturing systems are usually linked with CAM systems which are in turn linked to material handling systems. This is able to control parts and material flow. The main advantages of flexible manufacturing are better productivity, quicker machinery set-up times, lower labour costs and reduced down time on machinery. As a result, production and productivity are increased.

Quick response manufacturing

The main aim of quick response manufacturing (QRM) is essentially to reduce the lead time in all aspects of the manufacturing process. From outside a company, QRM is seen by the client as responding rapidly in the designing and making of products that are customised to their needs. From the company's point of view, QRM is geared to the improvement of the quality of the final product and to the reduction of costs.

Unit 3.2: Manufacturing systems, where both FMS and QRM are dealt with in considerable depth.

Production control

In any manufacturing situation it is vital that all the processes undertaken are carefully and accurately controlled. This means that at various stages during manufacture the quality of the work is checked. If problems are found, relevant procedures need to be activated. In modern industrial situations production control is undertaken using computer technology. There are a number of ways in which production can be monitored and controlled and they depend very much on the sophistication of software available. All involve monitoring and testing at various stages during production.

One method of quality control (QC) uses coordinate measuring. In this case, a sensor connected to a computer comes in physical contact with the object under scrutiny. The probe is able to check various parts of the component. If the original coordinates of the probe are known, then as the probe is moved around the work, the dimensions can be checked for accuracy. One method of checking for accuracy is to have an optical based system. The advantage of such a system is that there is no direct contact with the work – large objects can be checked and the response time is only limited by the electronics in the system.

FACTFILE:

Advantages of computer-aided inspection methods

- Measurable quality throughout the entire process.
- Inspection time can be speeded up.
- Any issues can be instantly identified and rectified early.
- Large 3D objects can be tested.
- Can be made to adhere to up-to-date and relevant standards.
- Information can be transferred digitally instantly.

There are still instances when visual inspection of products needs to be carried out. The problem with visual testing is that an element of human error can creep in. Visual testing usually consists of a random inspection of a sample of completed components. Naturally it is too late in the process to correct errors and 'failures' have to be scrapped. One method of increasing the sampling in visual testing is to introduce so-called 'intelligent cameras'. These can be programmed to view selected elements of the component and flag up any issues electronically. The advantage of this system is that instead of sampling, the camera can check all components passing in front of it.

Marketing, distribution and retail

Electronic point of sale

Information lies at the centre of any business and, if used properly, it ensures the business stays one step ahead of its competitors. By using electronic point of sale (EPOS) systems, a business is able to supply and deliver its products and services faster by reducing the time between the placing of an order and the delivery of a product.

Each product is identified using its unique barcode. When passed over a barcode reader, or scanner, the barcode is read by a laser beam. The laser scans the barcode and reflects back onto a photoelectric cell. The bars are detected because they reflect less light than the background on which they are printed. Each product has its own unique 13-digit number. The first two numbers indicate where the product was made, the next five are the brand owner's number, the next five are given by the manufacturer to identify the type of product and the final digit is the check digit, which confirms that the whole number has been scanned correctly.

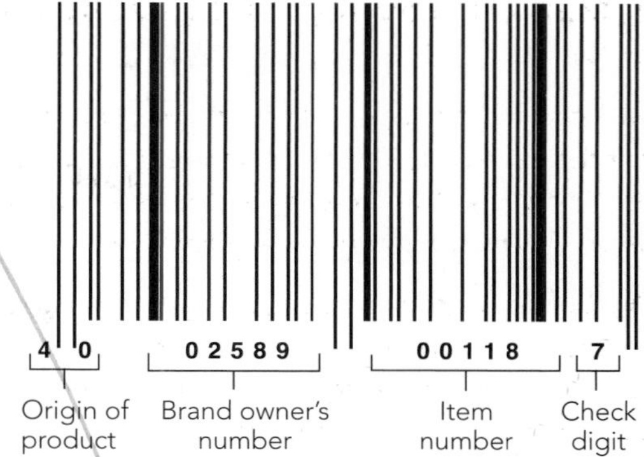

Figure 3.2 *A standard 13-digit barcode*

It is important to note that information regarding the price of the product is not contained on the barcode. Instead, the scanner (for example at a supermarket checkout) transmits the product code number to an in-store computer that relays the product's description and price back to the checkout, where it is displayed electronically and printed on the till receipt. The in-store computer then deducts the item purchased from the stock list so that it can be re-ordered when stock is low.

Data matrices, also known as 2D barcodes, are visual codes that can be read and decoded by machine vision systems. The increasing use of data matrix codes arises from the manufacturers' requirements for tracking their products or components. The intention is that batch or serial numbers can be permanently marked onto components, which is useful for tracking defective batches and identifying counterfeit parts.

Figure 3.3 *Data matrices (2D barcodes) are increasingly used as they can contain more information because they use not only the width of the lines but also the height of them*

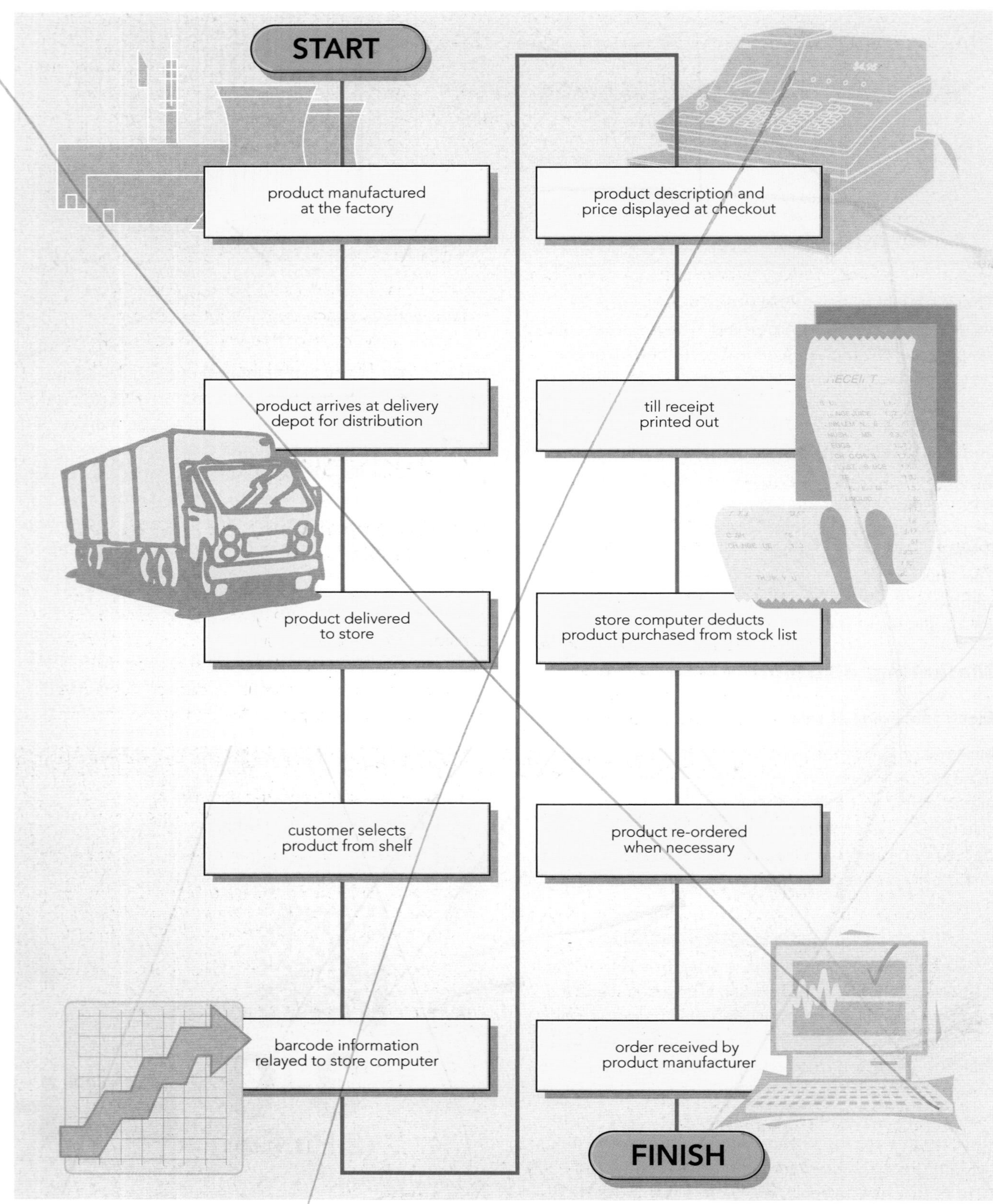
START
product manufactured at the factory
product arrives at delivery depot for distribution
product delivered to store
customer selects product from shelf
barcode information relayed to store computer
product description and price displayed at checkout
till receipt printed out
store computer deducts product purchased from stock list
product re-ordered when necessary
order received by product manufacturer
FINISH

Figure 3.4 *The electronic point of sale (EPOS) system*

Electronic point of sale and the associated management software provide manufacturers with:

- a full and immediate account of the financial transactions involving the company's products
- data that can be input into spreadsheets for sales/profit margin analysis
- the means to monitor the performance of all product lines, which is particularly important in mass production as it allows the company to react quickly to demand
- accurate information for identifying consumer buying trends when making marketing decisions
- a full and responsive stock control system by providing real-time stock updates
- a system that ensures sufficient stock is available to meet customer needs without over-stocking, which ties up capital.

Internet marketing and sales

The development of the Internet as a means of competing in a global marketplace has revolutionised the marketing and sales of products and services. Through the global networking of computers, the Internet provides an effective means of accessing a wealth of information and entertainment for anyone with a computer connected to the Internet. The dramatic rise in e-commerce has led to the formation of virtual communities, which businesses are eager to explore.

The possibilities for innovative marketing techniques are endless due to the simple identification of target market groups by user preferences. Marketing can be 'tailor made' to suit these markets, so a marketing message can be sent directly to potential customers as opposed to 'blanket' advertising in traditional media.

THINK ABOUT THIS! 

As a consumer, what are your main reasons for buying products over the Internet? Is it convenience, price or that you can buy products from other parts of the country or even the world?

Biotechnology

Genetic engineering in relation to wood production

In recent years, the subject of genetic engineering, or modification of living organisms, has generated considerable debate. In terms of timber production, genetic engineering can be used to change the properties of the wood that is produced. Nowadays, greater demands are put on the timber industry. For

Advantages	Disadvantages
To manufacturers and retailers • Worldwide reach and access to new markets and increased customer base • Increased company profile on a worldwide basis • Faster processing of orders and transactions, resulting in efficiency savings and reduced overheads • Detailed knowledge of user preferences and market trends by tracking sales • Cost savings due to reduced sales force and need for retail outlets • Less expensive than traditional advertising media such as television or magazines • Innovative marketing tactics can be employed that target specific groups **To the consumer** • Access to a wide range of products and services • Availability of product information to inform purchasing decisions • Online discounts and savings through price comparison websites • Convenience of shopping at home	**To manufacturers, retailers and the consumer** • Security concerns regarding input of personal bank details when purchasing goods • Personal information can be shared with other companies without customer consent • Difficult to find websites without exact details, resulting in a need for other expensive marketing methods, e.g. magazine adverts, Internet service provider sponsored searches, etc. • Slow Internet connection can cause difficulties in accessing information • Difficulty in navigating complicated web pages • Does not allow 'hands on' experience of product, i.e. touch, taste and fit • Access to inappropriate material • Spread of junk mail and threat of computer viruses

Table 3.9 *Advantages and disadvantages of Internet sales and marketing*

example, more and more homes are being built and in those new buildings, wood is required for roof timber and trusses. Timber is being used at an ever increasing rate and, as the UK has only limited managed forested area, much of that timber is being imported.

Biotechnology is at the forefront of research and experimentation. Trials have been undertaken to genetically modify (GM) trees to increase the amount of timber being produced. GM trees are the result of gene manipulation. This involves artificially inserting a gene from one plant into another, producing a change in the tree's biological characteristics.

The main advantage of GM trees to the paper and board industry is that trees can grow faster so that supply can keep up with demand. Other advantages include producing trees that are resistant to disease and insect attack to provide high-quality products and, in some instances, to provide specifically coloured timbers that can be marketed with specific aesthetic qualities.

GM tree technology is gathering pace with field trials rapidly increasing around the world. Overall genetic modification activities in forestry are taking place in at least 35 countries worldwide and are likely to make their commercial debut in Chile, China and Indonesia. There have been several trials in the UK so far, including GM elm trees resistant to Dutch elm disease, which were successfully grown in Scotland.

Advantages	Disadvantages
• Aids resistance of trees to disease. • Produces trees with increased growth rate. • Better forest management, which reduces deforestation.	• Long-term side effects not yet apparent. • 'Escape' of modified genes into natural ecosystems. • Development of tolerance to the modified trait by insects or disease organisms. • Rapid growth could cause shorter, more intensive rotations, resulting in greater water demand and reduced opportunity for nutrient recycling.

Table 3.10 *Advantages and disadvantages of genetically modified (GM) trees*

WEBLINKS:

www.arborea.ulaval.ca/faq/index.html#488

This website contains some interesting questions and answers regarding the genetic modification of trees.

THINK ABOUT THIS!

Scientists are artificially manufacturing new crops, materials and animals. How does this make you feel? The public reaction to GM crops and livestock has been quite negative – but what about the advantages? Think about the benefits to industry and the potential to solve problems in the developing world.

Micro-organisms to aid disposal of environmentally friendly plastics

Biodegradable polymers

Biodegradable polymers are materials derived from renewable raw materials that decompose in the natural environment. Biodegradation of polymers is achieved by enabling micro-organisms in the environment to break down the molecular structure of the polymer to produce an inert material that is less harmful to the environment.

Many biodegradable polymers, such as polyhydroxyalkanoate (PHA), are fully biodegradable as they are derived purely from renewable sources. Other types are semi-biodegradable, mixing renewable sources with existing petroleum-derived synthetic polymers. At present, the use of biodegradable polymers is in its infancy. Once their production increases they can become more economically viable in relation to synthetic polymers and therefore provide a widespread substitute.

Advantages	Disadvantages	Applications
• Fully degradable in suitable conditions, e.g. sun, moisture and oxygen. • Reduction of time in landfill and the associated harmful effects. • Starch-based plastics are formed from carbon that is already in the eco-system so does not contribute to global warming.	• Degradation of some plastics still contributes to global warming through the release of carbon dioxide as a main end product. • Damages recycled plastics when mixed and reduces its value. • Fully biodegradable polymers are more expensive as they are not widely produced to achieve large economies of scale. • May not be as energy efficient to produce as synthetic polymers, e.g. polypropylene. • Semi-biodegradable polymers remain in the environment for years.	• Packaging, e.g. blow moulded bottles. • Disposable products used in the food industry, e.g. utensils and dishes. • Plastic wrap for packaging, e.g. moisture barrier films for hygienic products. • Coatings for paper and board. • Agricultural uses, e.g. slow-release pesticides and fertilisers, mulches that degrade over time. • Medical uses, e.g. gauzes, sutures, implants. • Pharmaceutical uses, e.g. coatings for pills. • New natural fibres for the textiles industry.

Table 3.11 *Advantages, disadvantages and applications of biodegradable polymers*

Producing materials that are totally recyclable

One of the great issues of the twenty-first century is how we care for and protect the environment. Recycling is one of the ways in which the earth's natural resources can be preserved. All manufacturers claim that their particular product is recyclable in some way or another.

However, most materials, apart from metal, when recycled often have to be used for less demanding applications. For example plastics can be recycled but the end result of the recycling may be a less useful plastic. Woods can biodegrade so in one sense they can be reused in the form of composted down material. It could be argued that, apart from specially developed plastics such as Biopol®, the only truly totally recyclable material is metal. One example is aluminium, which is an element and as such is indestructible. When aluminium is recycled it does not degrade in any way so any aluminium that is salvaged and recycled can be used over and over again.

Almost all metals can be recycled into high-quality, usable metal.

Recycling metals is good for the environment and saves energy as it uses fewer natural resources, as well as reducing CO_2 emissions.

LINKS TO:

Unit 3.4 Sustainability: Minimising waste production.

THINK ABOUT THIS!

We are always being asked to recycle more of our materials. There are advantages to recycling in terms of not filling landfill sites and wasting finite resources. However, think about recycling from the other point of view. What might be the disadvantages to recycling?

WEBLINKS:

www.recyclemetals.org/

Biopol®

Biopol® is a trade name of the British chemical company ICI for the first fully biodegradable polymer, polyhydroxybutrate (PHB), developed commercially in the early 1990s. The first uses for Biopol® were in the packaging industry to produce blow moulded shampoo bottles as the material is water resistant and provides an effective barrier to air. 'Green' credit cards made from Biopol® have been introduced as a replacement for some of the 20 million credit cards in circulation today.

Biopol® is produced in nature through the fermentation by the *Alcaligenes eutrophus* bacteria of sugars (glucose) and carbohydrates that are collected in their cells as reserve material. Once this reserve material

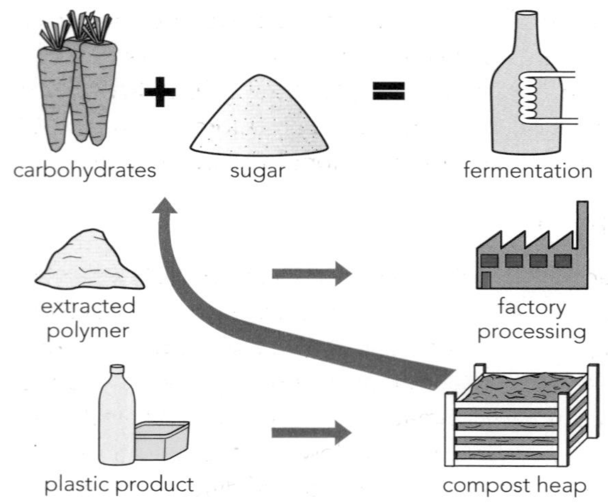

Figure 3.5 *The production of Biopol®*

is separated and refined from the bacteria, a white powdered polymer is extracted. This polymer can then be used in the usual manufacturing processes to produce plastic products. The significance of this is that Biopol® is produced naturally by renewable agricultural resources and, most importantly, it is fully biodegradable.

Biopol® is stable when stored in air and is quite stable when stored even in humid conditions. Degradation to carbon dioxide and water will occur only when the polymer is exposed to micro-organisms found naturally in soil, sewage, river bottoms and other similar environments. The rate of degradation is dependent on the material thickness and the number of bacteria present. Landfill simulations over a 19-week period show test bottles experienced a weight loss ranging from 30 per cent with oxygen present to 80 per cent with no oxygen present. The fact that Biopol® decomposes more rapidly without oxygen present is significant because oxygen is not present in modern sealed landfills.

Innocent™ eco-bottle

Innocent™ introduced an eco-bottle made from a material called polylactic acid (PLA), which is derived from corn starch. The use of PLA over regular polyethylene terephthalate (PET) and high-density polyethylene (HDPE) bottles has certain advantages:

- regular plastic bottles use finite resources (for example, oil and gas) whereas corn is a renewable resource
- PLA is made using a totally carbon-neutral process – no greenhouse gases are emitted in its production
- PLA is biodegradable, so it breaks down safely and relatively quickly.

Innocent™ encouraged its customers to compost these bottles with all of their other green waste (e.g. garden waste, food scraps). The eco-bottle can be recycled along with regular plastic bottles in the normal way but, to maintain the quality of the recycled plastic, compostable bottles should not make up more than 1 per cent of the mix.

Innocent™ admits that these corn starch bottles are a very new development so there is currently a lack of composting infrastructure around. Only 10 per cent of the UK's recycling centres have machines that can sort these bottles from the normal PET and HDPE plastic bottles. Therefore, the company has to be careful about how many compostable bottles it releases into the market before the correct facilities are in place to handle them.

WEBLINKS:

www.innocentdrinks.co.uk – eco-friendly drinks manufacturer.

LINKS TO:

Unit 3.4 Sustainability: The use of biotechnology is directly related to issues regarding sustainability.

Figure 3.6 *Innocent's eco-bottle is made entirely from a 100 per cent renewable source: corn. It is manufactured in a carbon-neutral process and is totally compostable, helping to reduce landfill waste*

Putting additives in plastics

Over the last fifty years the development of plastics has led to a revolution in the manufacture of products notably with new methods of manufacture and in some instances new and even 'smart' materials.

Plastics are manufactured using polymers which are often mixed with other ingredients. These ingredients, known as 'additives', are added to plastics to enhance, or change, their properties. Without additives, the use and applications of plastics would be somewhat limited. The use of additives makes plastics easier to process, enhances their aesthetics, makes them safer and even reduces their impact on the environment.

Plasticisers

There are occasions when, for particular products, the plastic being used needs to be soft and flexible. For example, the polyvinyl chloride (PVC) outer coating of an electrical cable needs to be relatively soft and flexible to allow the cable to be bent and shaped. In order to increase the flexibility of a plastic such as PVC, plasticisers are added. The most common plasticisers are organic compounds known as phthalates. Because they are organic compounds, they are biodegradable and break down relatively quickly.

Plasticisers are also added to assist in plastic manufacturing processes. In a process such as injection moulding it is important that the plastic flows evenly into every part of the mould. To help the plastic flow correctly, plasticisers are added during the manufacture of products.

Fillers

Sometimes it is necessary to increase the bulk of plastics. In order to do this fillers are added. Fillers are also used as an economic measure in order to reduce the overall cost of the plastics produced.

Fibres

Plastics on their own are sometimes not strong or stiff enough to carry out the particular task that is required of them. For instance, a golfer would not want a flexible golf club! A golf club needs to be strong and stiff. In order to create a plastic golf club that is both strong and stiff, carbon fibres are added to the plastic.

Carbon fibre reinforced plastics are composite materials that use a polymer matrix reinforced with thin carbon fibre strands. The plastic resin is usually epoxy or polyester. The resulting material is extremely strong and is widely used in the aerospace and automotive industries. The most common example of using fibres in plastics that is often seen in schools is glass reinforced plastic (GRP), where the fibres are in the form of glass matting.

Stabilisers

Over a period of time polymers can deteriorate. For example, where polymers are exposed to ultraviolet light, reactions can occur that cause the polymer to degrade. In order to help retard degradation, stabilisers are added to the material. A stabiliser is an additive that helps to prevent, or slow down, the degradation of polymers. There are number of different types of stabilisers used in plastics, all of which have particular roles to play in the material.

- **Heat stabilisers**
  These are put into the plastic to prevent it from decomposing during any manufacturing process that uses heat (such as injection moulding). During the injection moulding process the heat of the plastic can reach temperatures in excess of 180° and without stabilisers the plastic might well disintegrate.
- **Light stabilisers**
  When plastics are exposed to sunlight, the ultraviolet light in the sunshine can make the plastic go brittle and break down. This is sometimes seen in uPVC window frames. After a number of years in sunlight, uPVC can break down and crack. Light stabilisers are added to plastics in order to slow down chemical degradation from exposure to ultraviolet light.
- **Biostabilisers**
  In some cases plastics can be attacked and broken down because of microbiological attacks. In such cases, plastic can become stained and discoloured and even begin to smell, with a loss of aesthetics. However, the greatest problem with microbiological attack is that the plastic loses its mechanical properties. Biostabilisers are added to reduce the impact of microbiological attack.

Foamants

When a foamant is added to a plastic two things happen. Firstly, it increases the volume of the plastic and secondly, it increases the elasticity of the plastic, making it appear to have more of a 'spring'. This means that the plastic can be pushed or pressed and it will spring back to its original shape.

Plastics that contain foamants are also used as buoyancy aids in swimming. The cell structure of a plastic with foamants added is such that air gets trapped within the cells of the material, thus making the plastic lightweight and buoyant. Other applications include both heat and sound insulation.

WEBLINKS:

www.phthalates.com

This website gives some interesting information about additives in plastics.

Lamination

Lamination is where a material has been produced by gluing together thin sheets, or veneers, to make up that material. The most common form of laminate material is plywood. Plywood is made up of a number of thin layers of wood that are called veneers. There are always an odd number of veneers and they are always arranged so that the grain on each layer is positioned at 90° to the layer above and below. Because the veneers are arranged in this way, the resulting plywood sheet is very stable and very strong. The layers are bonded together with strong glues, usually epoxy-resins.

There are a vast number of applications for plywood, from interior uses such as cupboards and furniture constructions to exterior uses including boat building. Other wood laminates include blockboard and laminboard. Lamination is also used to give a protective surface to sheet material. For example in the kitchen a work surface that can be easily cleaned and kept clean is required. For this a laminate material such as Formica® is glued onto the sheet, usually using contact adhesives.

Lamination can also be used for shaping material. This is where sheets, or strips of veneer, are glued together and clamped in a former. When the glue dries and the work is removed from the former the work retains the shape of the former. This process is used in the production of products such as chairs where sweeping curves might be required. The advantage of this type of production is that it is much less time consuming than trying to bend a solid piece of wood into a desired shape. It also means that, once the former has been created, it can be used repeatedly and can be used as a jig to batch produce the product.

Advantages	Disadvantages
Economic to use • Laminated board uses the whole of the tree • The whole sheet can be used. There is little waste **Can be used as a basic material that can be veneered** • Can be used in flat pack furniture and then veneered to give good finish **Strong material** • It is strong in all directions **Comes in large sheets** • Sheets can be cut easily using jigsaw • Little wastage **Can be shaped into curve** • By using a former can be shaped into complex curves	**Surface finish not good** • The surface of laminated wood is often rough and will not always take paint or varnish **Edges not always good** • The layers of material are visible **Visible flaws** • Finished material often has evidence of knots and can be visually unappealing **Delamination** • If material becomes wet or damp layers may start to come apart

Table 3.12 *Advantages and disadvantages of using lamination*

THINK ABOUT THIS! 

Find an example of a piece of furniture that has been manufactured using lamination. Draw a sketch of what the former might have looked like that was used to produce the laminated shape.

LINKS TO:

Unit 2.1: Laminates Plywood and blockboard.

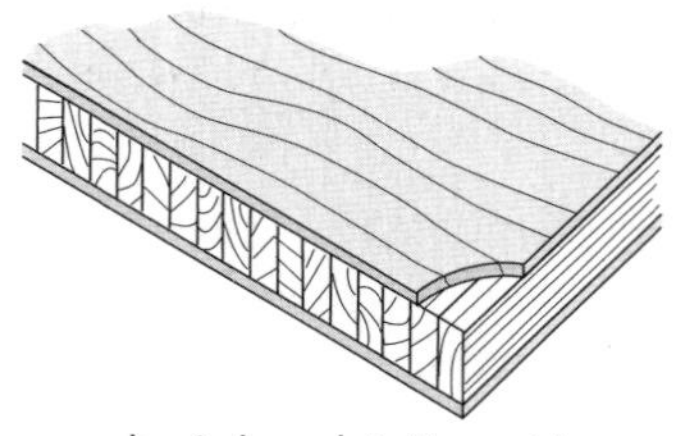

lamin board, 5–7mm strips

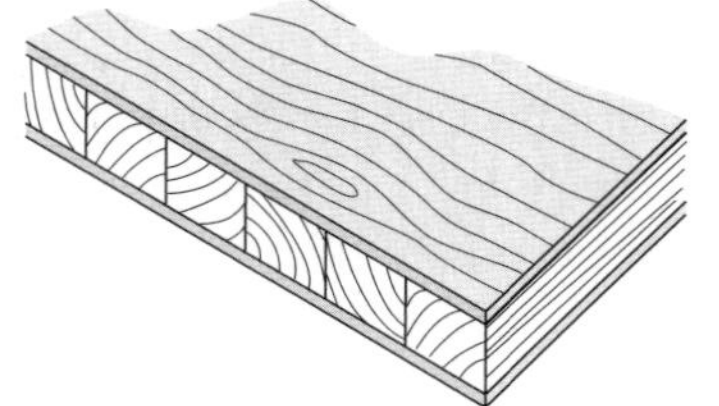

block board, up to 25mm strips

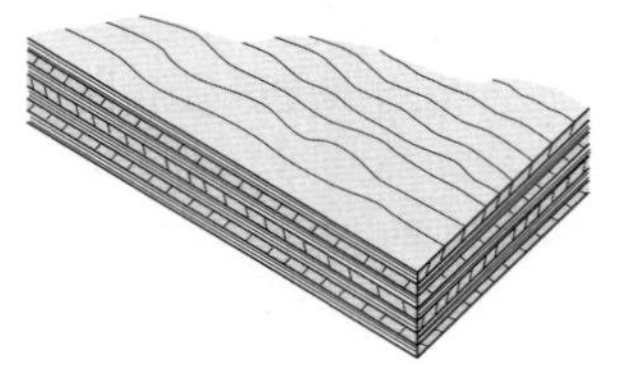

7-ply

Figure 3.7 *Laminated form of manufactured sheet timber*

Systems and control

Getting started

We are in a period of rapid development of manufacturing technologies, with many manufacturers now using complex automated systems to produce their products. We notice this in the shops with a wide range of products on sale, especially electronics, at prices that seem to be getting cheaper all the time. As modern designers we must understand these systems. But what effect will they have upon employment and our lives in the future?

Manufacturing systems

Advanced manufacturing technology

Advanced manufacturing technology (AMT) describes the significant impact of computers on manufacturing. Computer technology has led to the development of computer-aided design (CAD)/ computer-aided manufacture (CAM), robotics, materials-handling devices and computer-integrated manufacture (CIM) systems that have increased the accuracy and flexibility of the manufacturing process. AMT has also had an impact on manufacturing management with the integration of product development and manufacturing stages through the use of ICT.

Computer technology features strongly in automation used in modern manufacturing systems. Although automation preceded computer technology, it is computer power that has enabled automation to become flexible and more effective in applications other than mass production.

Quick response manufacturing

Quick response manufacturing (QRM) was developed to give companies a competitive advantage by increasing their operating efficiency. It requires a manufacturer to move from traditional batch production to 'flow' production. This flow production is triggered by consumer demand and not based on complex demand forecasts. By operating this system, a manufacturer can quickly respond to fluctuations in the economy and ever-changing market demands as well as avoiding excess stockpiling of products, which costs money.

QRM involves several concepts, such as total quality management (TQM), JIT and manufacturing cells, but its main aim is to increase the overall flexibility and responsiveness of the company. By, for example, manufacturing in cells, production teams can be dedicated to specific product lines. These teams can be quickly and efficiently reallocated if the product mix should change. Therefore, a manufacturer has increased production flexibility and will be better equipped to meet changing market demands. This is a more efficient arrangement than mass-production systems that are designed to take advantage of economies of scale by producing large batches, which encourages stockpiling.

In the ideal QRM situation, the manufacturer would begin production as soon as an order is initiated, suppliers would deliver raw materials directly to the production line, the product would be manufactured and the finished goods would flow directly to a waiting truck for delivery. Therefore, the primary functions of QRM are to pull raw materials through the production process strictly according to market demands and to ensure that every product component and order moves as quickly as possible throughout the entire supply chain.

Advantages	Disadvantages
• Lower working capital required as raw material and finished goods stocks are minimised, placing the manufacturer in a better strategic position to utilise resources and capital. • Better position to increase market share as quicker response times will attract new clients and allow the manufacturer to capture a larger percentage of the market. • Increased turnover of stock as production systems are triggered by demand, smaller batches are produced and stock decreases, resulting in lower storage costs. • Reduces the cost of quality by minimising waste and by placing more responsibility and accountability on specific production teams.	• Increased reliance on suppliers in order to react to demand and quickly accommodate production schedules. Poor supply could result in a manufacturer's inability to meet customer requirements, excessive stock and backorders. • Large variations in demand could cause problems if the manufacturer cannot react to the high production volume efficiently. • Managing and implementing the change required can be very difficult as QRM changes the roles and responsibilities of employees, which may cause friction. • To successfully implement QRM, a manufacturer must have representation and backing from all disciplines including production, planning, purchasing, engineering, manufacturing, quality, finance and human resources to facilitate the implementation.

Table 3.13 *Advantages and disadvantages to manufacturers of using quick response manufacturing (QRM)*

Concurrent manufacturing

In order to remain competitive and cope with increasing market pressure from mounting customer demands, manufacturers need to get to market first with products that customers want. Concurrent manufacturing provides a systematic approach to the integration of design, manufacture and related processes where all life cycle stages of the product are considered simultaneously. One of the main advantages of a concurrent manufacturing system is that it produces designs 'right first time' through the use of multi-disciplinary design teams, reduces product development times and enables the earlier release of new products.

The problems with product development performance that a concurrent manufacturing system aims to overcome are those of the traditional sequential product development process in which people from

Traditional sequential product development process:

Stage	Time to market (Weeks)											
	1	2	3	4	5	6	7	8	9	10	11	12
Project planning												
Design & development												
Materials supply & control												
Manufacture												
Delivery												

Development process using a concurrent manufacturing system:

Stage	Time to market (Weeks)											
	1	2	3	4	5	6	7	8	9	10	11	12
Project planning												
Design & development												
Materials supply & control												
Manufacture												
Delivery												

Figure 3.8 *Stages in concurrent manufacturing overlap significantly using multi-disciplinary design teams*

different departments work one after the other on successive phases of development. In traditional sequential development, the product is first completely defined by the design department, after which the manufacturing process is defined by the manufacturing department, etc. Usually this is a slow, costly and low-quality approach, leading to a lot of design changes, production problems, product introduction delays and a product that is less competitive than desired.

A concurrent manufacturing system, however, brings together members from a wide range of disciplines such as manufacturing, project management, technical support, marketing and other specialist areas who are integrated with the designers to form a multi-disciplinary team. Multi-disciplinary teams acting together early in the workflow can take decisions relating to product, process, cost and quality issues. They can make trade-offs between design features, part manufacturability, assembly requirements, material needs, reliability issues, serviceability requirements, and cost and time constraints. Differences are more easily dealt with early in design, making the design process quicker and more efficient as, ideally, no re-design is necessary and manufacture can start earlier.

One of the most important factors in any successful concurrent system is the effectiveness of the project team. It must function efficiently with few internal problems and a comprehensive understanding of its objectives. Therefore, excellent channels of communication are needed to be truly effective.

Computer-based systems enable efficient communication between individual team members and integrated project teams for product development. For manufacturing companies operating at a national, or international, level a computer network is essential to support data transfer between team members. Combining concurrent manufacturing systems, effective management and teamwork ensures the development of a high-quality and reliable product with low life-cycle costs in the shortest development time.

Early in the development stage, designers can use quality function deployment (QFD), a strategy for remaining 'in touch' with customer requirements in order to create a more successful product. Lead time has proved to be a significant aspect of modern competition. By decreasing the lead time, a company is able to respond rapidly to changes in market trends or to incorporate new technologies. By employing concurrent manufacturing systems, lead times can be noticeably reduced, creating a market advantage for those companies who are able to produce products rapidly.

QFD is a quality assurance method that factors customer satisfaction into the development of a product before it is manufactured. The main features of QFD are its focus on customer requirements, the use of multi-disciplinary teamwork and a comprehensive 'House of Quality' matrix. This matrix is used by the team to translate customer requirements into a number of prioritised targets to be met by the new product design. Some of the advantages of using QFD as part of a concurrent manufacturing system are:

- reduced time to market
- reduction in design modifications
- decreased design and manufacturing costs
- improved product quality
- enhanced customer satisfaction.

The six major components of the 'House of Quality' matrix

Component		Content
1	Customer requirements	A structured list of requirements taken from customer statements in a market survey.
2	Technical requirements	A structured set of relevant and measurable product characteristics.
3	Planning matrix	Illustrates customer perceptions observed in the market survey. Includes relative importance of customer requirements, and company and competitor performance in meeting these requirements.
4	Interrelationship matrix	Explores the strength of the relationship between the customer requirements and the technical requirements.
5	Technical correlation	Used to identify where technical requirements support or impede each other in the product design.
6	Technical priorities, benchmarks and targets	Quality targets against each technical specification point can be measured.

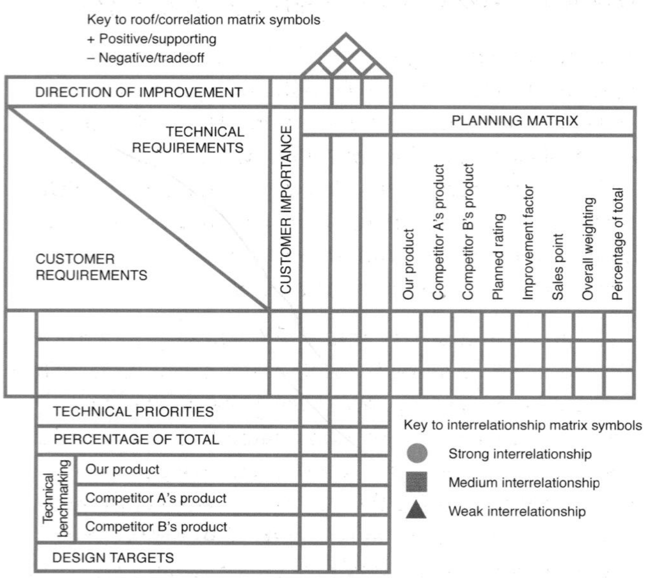

Figure 3.9 An example of a House of Quality matrix for assuring the quality of a design

www.gsm.mq.edu.au/cmit/qfd-hoq-tutorial.swf

Gives an animated explanation of a House of Quality matrix.

Flexible manufacturing systems

A flexible manufacturing system (FMS) is a form of flexible automation in which several machines are linked together by a material-handling system, with all aspects of the system controlled by a central computer. An FMS brings together new manufacturing technologies such as computer numerically controlled (CNC), robotics, automated material handling, etc. to form an integrated system. It is different from an automated production line because of its ability to process more than one product style simultaneously.

FMS have powerful computing capacities that give them the ability not only to control and coordinate the individual equipment items and facilities, but also to perform production planning and routing of material through the system. The main advantage of an FMS is its high flexibility in managing manufacturing resources in terms of both machines and personnel in order to manufacture a new product. This flexibility allows the system to react quickly to changes in production, whether predicted or unpredicted, utilising two main features.

- **Machine flexibility** – involves the system's ability to be changed to produce new product types, and ability to change the order of operations executed on a part.
- **Routing flexibility** – involves the ability to use multiple machines to perform the same operation on a part, as well as the system's ability to absorb large-scale changes, such as in volume, capacity or capability.

FMS vary in their complexity and size. Some are designed to be very flexible and to produce a wide variety of parts in very small batches. Others have the ability to produce a single complete product in large batches from a sequence of many individual operations known as a flexible transfer line.

The advantages of FMS include:

- increased productivity due to automation
- shorter lead times for new products due to flexibility
- lower labour costs due to automation
- improved production quality due to automation.

The main disadvantages of flexible manufacturing systems are that setting up the system requires a great deal of pre-planning and the very high set-up costs.

Computer-integrated manufacture

Computer-integrated manufacture (CIM) takes the concept of integration of separate manufacturing technologies developed by FMS a step further by bringing together all aspects of a company's operations, not just those that are directly involved in manufacture. Under a CIM system, all teams can share the same information and easily communicate with one another. A CIM system uses computer networks to integrate the processing of production and business information with manufacturing operations to create cooperative and smooth-running production lines. The tasks performed within CIM will include:

- design of the product using CAD
- planning the most cost-effective workflow
- controlling the operations of machines and equipment needed to manufacture the product
- performing business functions such as ordering stock and materials and invoicing customers.

Sub-systems found in a CIM operation

Manufacturing systems	Business systems
• CAD, CAM and CNC machines • FMS • Automated storage and retrieval systems (ASRS) • Automated guided vehicles (AGV) • Robotics	• Computer-aided process planning (CAPP) • Enterprise resource planning (ERP) • Product data management (PDM) software allowing computerised scheduling and production control • Computer-aided quality assurance (CAQ) • An electronic data interchange (EDI) business system integrated by a common database

One of the drawbacks with a CIM system is its dependence upon computer data to integrate fully all operations. For example the software from one brand of equipment may be incompatible or cause difficulties when integrated with CNC machinery and an automated storage and retrieval system (ASRS) from another brand. The cost of managing data is also a key issue within CIM. This is because if data becomes corrupted it may cause machinery to malfunction. In order to prevent this, companies often use a product data management system.

Product data management systems

Product data management (PDM) is an information system used to manage the data for a product as it passes from design to manufacture. The data includes plans, 3D models, CAD drawings, CNC programs as well as all related project data and documents. A PDM also manages the inter-relationships between the data so that when changes are made to one database, the effects are highlighted in the others.

PDM systems support production planning by the development, distribution and use of product data. The focus is on managing and tracking the creation, change and archive of all information related to a product. For instance the design of a component may go through many changes during the course of its development, each involving modifications to the CAD data. Once the designer is satisfied with the component, the PDM system may notify an analyst that the design is ready for them to perform a stress analysis on it. When that task is complete, the stress analyst performs an electronic sign-off. The PDM system will then notify a manufacturing engineer that the component is ready for planning of its manufacture and a tool designer that the component design is ready for a tool design. After these tasks are performed, the various team members submit their data to the PDM system for a final review and team sign-off, after which it is released to begin manufacture.

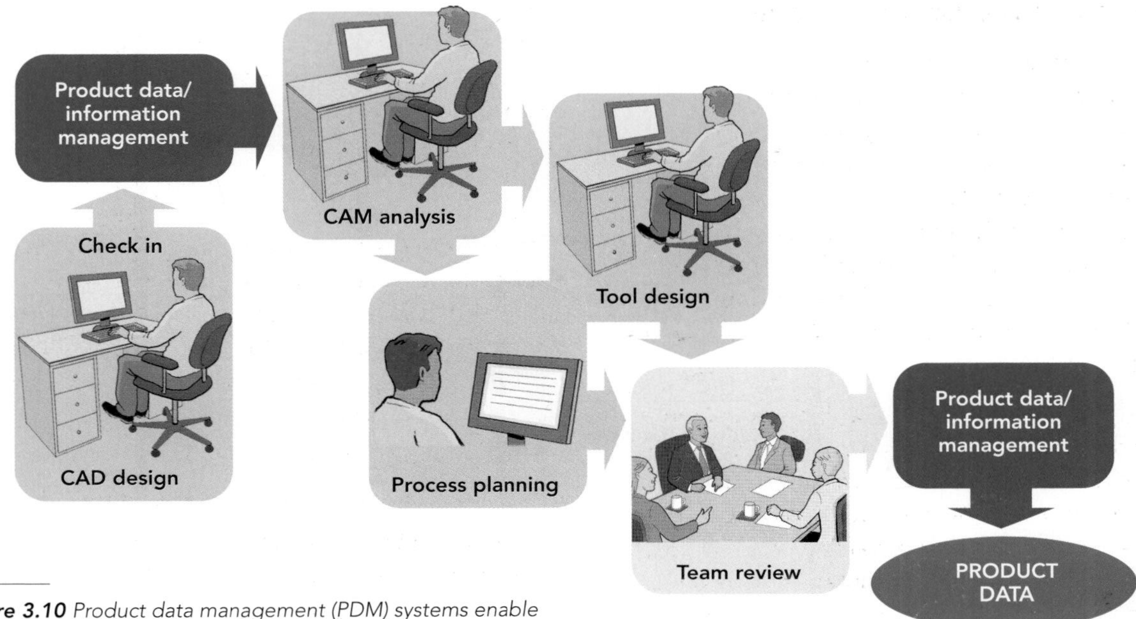

Figure 3.10 *Product data management (PDM) systems enable efficient collaboration when planning production*

The advantages of a product data management (PDM) system include:

- **reduced time-to-market** as data is instantly available to all teams for review, eliminating 'bottlenecks' where documents await distribution or sign-off
- **improved productivity** as changes to the product data are tracked and managed automatically, reducing the time taken to search and retrieve documents and giving the ability to reuse design data without repeating work
- **improved control** due to efficient management systems that assure everyone is working from the most current data; access control features ensure that only authorised team members can access or change shared information.

Enterprise resource planning systems

Enterprise resource planning (ERP) systems attempt to integrate all departments and data across a company onto a single computer system that can serve all those different departments' particular needs by using a unified database. For example, the finance, human resources, manufacturing and warehouse departments still use their own software but ERP combines them all together into a single, integrated software program that runs off a single database so that the various departments can more easily share information and communicate with each other.

ERP improves the way in which a company takes a customer order and processes it into an invoice and revenue. ERP takes a customer order and provides a software 'road map' for automating the different steps along the path to fulfilling it. An example of such a road map is when a customer order is entered into an ERP system; all the information necessary to complete the order is instantly accessible, such as the customer's credit rating and order history from the finance department, stock levels from the warehouse department and the delivery schedule from the logistics department. Employees in the different departments all see the same information and can update it instantly. When one department finishes with the order it is automatically routed via the ERP system to the next department. Therefore, any order can be easily tracked and customers should receive delivery of their orders faster and without errors.

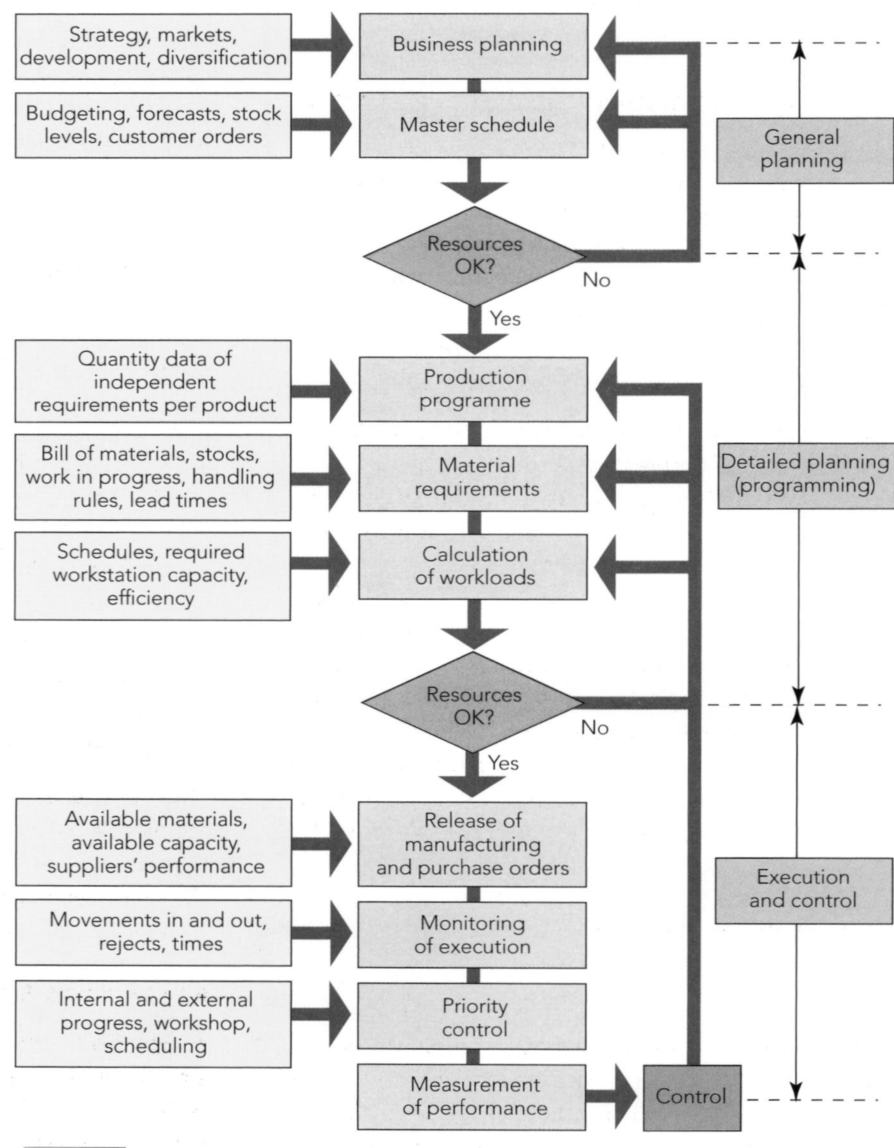

Figure 3.11 *Enterprise resource planning (ERP) in a modern manufacturing system*

ERP systems are extremely expensive to install and costs are incurred during the 'switching over' period. The main problems with ERP systems, however, often arise from issues relating to the training of the workforce. To get the most from the system, employees must adapt to the new work methods as prescribed by the software. ERP systems can be difficult to use and considerable training is required if all departments are to effectively implement the change. Resistance to change is always a common reason for failure of systems in business and industry and some departments may not want to share sensitive information, or a weak department could compromise the efficiency of the whole system. Therefore, success depends on the skill and experience of the workforce and ongoing training.

Lean manufacturing and JIT systems

Lean manufacturing is, as the name suggests, a manufacturing where there is no 'fat'. A key feature of lean manufacturing is the notion of 'just in time' (JIT) – that is there are no warehouses full of materials waiting to be used. Materials arrive just when they are needed. The objective of lean manufacturing, then, is to provide techniques that ensure minimum waste is incurred during production and to produce products only when they are needed.

JIT is derived from a Japanese manufacturing philosophy. Quite simply, JIT ensures that the right materials, components and products arrive at the right time, at the right place and in the exact amount. This reduces waste and overstocking as new stock is only ordered when it is needed, so saving warehouse space and storage costs. This focus on producing the right amount at the right time relies upon accurate analysis and forecasts using the right information at the right time. However, if a manufacturer is inaccurate in their predictions, such as a rise in demand for the product, stock will be used up rapidly with very little opportunity for re-supply. Production can also be held up or shut down completely if the raw materials supplier has problems fulfilling orders.

Pull tools – Kanban

A lean organisation needs to be tailored so that orders are pulled through the production system. A Kanban system uses cards, or containers, as simple visual signals to indicate when to pull materials, components or products through the production system. The system relies on a simple rule of only producing or delivering when a card or empty container is passed to a workstation or manufacturing cell. An integrated

Key stage		Content
1	Value	• Focus on value in the context of what the customer/end-user is prepared to pay for
2	Value stream	• Identify how value-adding and non-value-adding activities affect efficiency throughout the production of the product • Examples include: for value-adding activities: machining, processing, painting, assembling, etc. for non-value-adding activities: scrapping, sorting, storing, counting, moving, etc.
3	Flow	• Design processes that result in uninterrupted material flow, from raw materials to delivery of finished product
4	Pull	• Design manufacturing for 'pull' of product through the process as a response to demand, rather than 'push' of raw materials into the process by producing components irrespective of whether they are needed or not
5	Perfection	• Adopt an approach that continually improves working processes by adding greater value and by reducing waste

Table 3.14 *The five key stages of lean manufacturing*

Kanban involves production and transportation systems, the difference between the two being the information contained on the card or container:

- production Kanban includes details of the operations that need to be carried out at the workstation or manufacturing cell
- transportation Kanban only contains details regarding where the materials, components or products have come from and where they are going to.

The main benefit of using Kanban is that it reduces the amount of work in progress and finished goods in stock. Kanban restricts the supply of materials and components until they are needed, which provides an effective JIT system.

Perfection tools – Kaizen

Kaizen is also known as continuous improvement, where small changes are made to the production process resulting in small improvements being made. This system tends to be carried out on a regular basis as the changes made are often low cost and the improvements made tend to be small.

Flexible manufacturing cells

As customers demand variety and customisation of products as well as specific quantities delivered at specific times, a lean manufacturer must remain flexible enough to serve its customers' needs. Manufacturing cells allow manufacturers to provide their customers with the right product at the right time. It achieves this by grouping similar products into families that can be processed on the same equipment in the same sequence.

A manufacturing cell is a group of workstations, machines or equipment arranged such that a product can be processed progressively from one workstation to another without having to wait for a batch to be completed and without additional handling between operations. Cells enable the production process to be broken down into discrete segments or modules and can be dedicated to a process, a component or an entire product.

Modular production methods provide flexibility as upgrades to processes can be performed relatively quickly and easily by shutting down one cell whilst another simultaneously opens and takes its place. This means that the entire production line does not have to be disrupted, which reduces costs from stoppages, or 'downtime'. Flexibility also occurs when a large variety

Types of manufacturing cells

Functional cells	These cells perform a specific function as opposed to manufacturing a complete product and consist of similar equipment. For example, a factory that primarily carries out machining operations may have a bank of lathes together in a 'turning cell'. This type of cell is not as flexible and can produce higher levels of waste.
Group technology (or mixed model) cells	These cells perform a series of operations for several different product lines. These products often involve very similar manufacturing operations though not usually identical. This type of cell can work very well within a lean manufacturing environment, particularly if the company is characterised by a large product range with low volume.
Product focused cells	These cells are product focused and typically manufacture one type of product through a series of operations. These are the ideal lean manufacturing cell and are perfect for a small product range with high volume manufacturers.

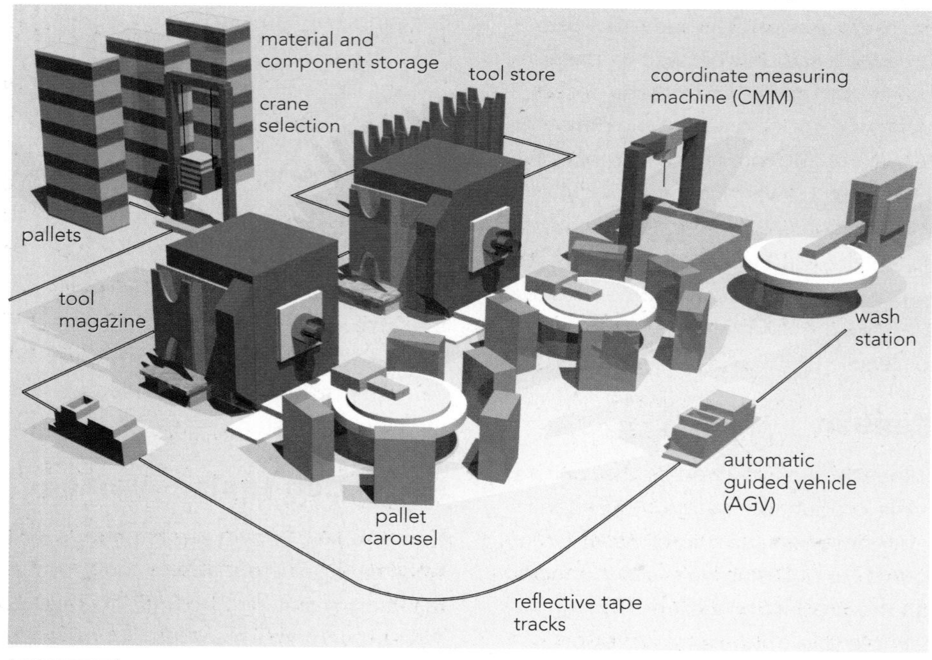

Figure 3.12 *A fully automated manufacturing cell*

Advantages	Disadvantages
• Greater control of the production process enabling fully automated production • Safer working environments due to removal of risks to humans • Flexible production as CNC machinery can be quickly reprogrammed to manufacture alternative components, reducing set-up times • Scale of production can be directly linked to customer demand, responding quickly to changes in quantities • Improved productivity as production rates are consistent, less waste is generated and production costs are reduced • Reduced manufacturing times as efficient cutting paths are generated by software • Increased operational reliability and consistency in repetitive tasks, maintaining the high quality of products produced	• Extremely high set-up costs, as expensive machinery and installation are required • Negative effects on employment as CAM requires less human involvement in its operation • Worker morale may be affected due to 'machine minding' job roles

Table 3.15 *Advantages and disadvantages of computer-aided manufacture (CAM)*

of products are assembled from a range of similar components. For example cellular manufacture enables a large variety of cars to be produced using various combinations of customer options such as engine sizes, interior finishes, etc. By combining the outputs from a number of work cells, each with a very narrow range of functions, it is still possible to take advantage of the mass production efficiencies of specialisation and scale.

Typically, a manufacturing cell involves 3–12 workers and 5–15 workstations in a compact arrangement. However, some manufacturing cells are fully automated using CAM containing several CNC machines, computer-aided quality control and automated materials-handling systems. This is an ideal cell layout as it manufactures a narrow range of very similar products and is self-contained with all necessary equipment and resources. Materials sit in an initial queue when they enter the cell. Once processing begins, they move directly from process to process, resulting in a very fast throughput.

Computer-aided quality control systems

Computer-aided quality (CAQ) control can be achieved within a manufacturing cell using a coordinate measuring machine (CMM) for extremely accurate dimensional measurement. A CMM is a mechanical system designed to move a measuring probe to determine the coordinates of points on the surface of a workpiece to measure the sizes and positions of features on mechanical parts. This provides data that can immediately be fed back into the production process to analyse extremely small tolerances and control the quality of components. Laser scanning technology is advancing rapidly, where many thousands of points can be taken and used to not only check size and position, but to create a 3D image of the part as well.

In addition, other systems can provide automatic identification such as optical character recognition and barcode readers. In the food industry, a high-speed product identification system is designed to read a special code on every can of food at 1,200 cans per minute. This prevents accidental product mixing and wrongly labelled products reaching customers.

Automated materials-handling systems

Materials-handling systems provide transportation and storage of materials, components and assemblies. Materials-handling activities start with the unloading of goods from delivery transportation; the goods then pass through storage, onto machining, assembly, testing, packing and finally loading onto transport. Each of these stages of the production process requires a slightly different design of handling equipment, for example unloading from a lorry may require the manual operation of a fork-lift truck.

However, CIM systems enable a range of automated materials-handling systems to operate within the workplace. Fully automated handling systems ensure that the materials, components and assemblies are delivered to the production line when required without significant manual intervention. The one downside of these systems is that, as with all automation, there are repercussions for the workforce in terms of downsizing the number of workers required in a factory.

Figure 3.13 An automated storage and retrieval system (ASRS) used in a warehouse

Automated storage and retrieval systems

An automated storage and retrieval system (ASRS) is an automated robotic system for sorting, storing and retrieving items in a warehouse. Within CIM, a computer controls the transportation of materials and components to the required points. All stocks of materials and components are stored in racking systems. The ASRS system will select the correct component from the rack, retrieve it by means of a crane and place it on a conveyor, or onto an automated guided vehicle (AGV), for transportation.

Warehousing and distribution are about moving as much product to market as possible, in the shortest amount of time. While there are a variety of methods available, it is becoming increasingly necessary for companies of all sizes to replace manual methods with innovative automated storage and retrieval systems. Traditional manual systems have a limited throughput of approximately ten load movements per hour, which is in stark contrast to the average of 40 load movements per hour that are performed by an ASRS. In addition to increasing throughput and reducing labour costs, employees' technical skills are developed in the operation of such a system.

Automated guided vehicles

An automated (or automatic) guided vehicle (AGV) is a materials-handling device that is used to move parts between machines or work-centres. They are small, independently powered vehicles that are usually guided by radio frequency wires that are buried in the floor, or use optical sensors in a laser-guided navigation system. They are controlled by receiving instructions either from a central computer or from their own on-board computer.

www.amhsa.co.uk/case_studies.htm

Case studies from the Automated Material Handling Systems Association.

AGV type	Application
Towing vehicles	• These were the first type introduced and are still a very popular type today to pull a variety of trailer types
Unit load vehicles	• These are equipped with decks that permit unit load transportation and often automatic load transfer • Decks can either be lifted and lowered type, powered or non-powered roller, chain, belt decks or custom decks with multiple compartments
Pallet trucks	• These are designed to transport palletised loads to and from floor level, eliminating the need for fixed load stands
Fork truck	• These have the ability to service loads both at floor level and on stands • In some cases these vehicles can also stack loads in a racking system
Light load	• These are used to transport small parts, baskets or other light loads through a light manufacturing environment and are designed to operate in areas with limited space
Assembly line vehicles	• These are an adaptation of the light load AGVs for applications involving serial assembly processes such as manufacturing cells

Table 3.16 *Types and applications of automated guided vehicles (AGVs)*

The impact of advanced manufacturing technologies on employment

As with any computer-aided technology, such as CAM, there has been some reduction of workforces as machines have become increasingly efficient. However, it does not eliminate the need for skilled professionals such as creative product designers, manufacturing engineers and CNC machine programmers. Computers, in fact, increase the levels of skill in many manufacturing professionals who use advanced productivity, visualisation and simulation tools.

The Manufacturing Institute in the UK argues that the media magnify the negative aspects of manufacturing. It outlines the 'Top ten manufacturing myths' in order to promote manufacturing in a more positive light.

Myth 01: Manufacturing jobs are all monotonous, strenuous and low paid

To the majority, the image most vividly conjured up when thinking of manufacturing is still one of endless assembly lines; employing poorly paid manual workers who carry out the same mundane tasks, whilst working in the same metre space, year upon year.

However, the Manufacturing Institute argues that:

Over the past 20 years production lines have become increasingly automated, leading manufacturers to demand increased skills flexibility among their staff... Employers now require a multi-skilled workforce to service an increasingly challenging, diverse and multi-faceted industry...with latest figures confirming that manufacturing jobs do in fact compare favourably with those in the service sectors – including banking, retail and the creative industries.

Myth 08: Manufacturing is not a creative industry

The case against this myth is simple – everything that is manufactured has to be designed...All components need to be functionally and creatively designed for purpose. A good functional design can save a company millions in production costs by eliminating waste, stock surplus and lead times, whilst a creatively designed unit will also help sell a product.

THINK ABOUT THIS! 

Manufacturing industries are increasingly located in countries such as China rather than the UK. Why do you think this is? What could be done in the UK to promote careers in the manufacturing industries?

WEBLINKS:

www.manufacturinginstitute.co.uk

The Manufacturing Institute is an advisory service to manufacturers and universities in the North West.

Robotics and artificial intelligence

Robots in automated manufacturing systems

The vast majority of robots in use today are found in the manufacturing industry on automated production and assembly lines and in manufacturing cells. Automation is the use of computer systems to control industrial machinery and processes, largely replacing human operators. The British Automation and Robot Association defines an industrial robot as:

A re-programmable device designed to both manipulate and transport parts, tools, or specialised manufacturing implements through variable programmed motions for the performance of specific manufacturing tasks.

Japan is the world leader in robotics technology and they widen this definition to include arms controlled directly by humans which have a wide range of possible future applications.

The basic robotics technology in modern industrial robots is similar to CNC technology but most robots have many degrees of freedom. In manufacturing applications, robots can be used for assembly work, processes such as painting, welding, etc. and for materials handling. More recently robots have been equipped with sensory feedback through vision and tactile sensors. In the future robots may also link more intelligently with humans so that they can judge for themselves when it is safe to operate without having built-in guards and safety mechanisms that limit their operations.

Levels of complexity of manufacturing tasks for robots

Level (Increase in difficulty ↓)		Applications
	1	Applications that can be achieved using a simple robot using jigs and fixtures to position components and tooling to achieve the required accuracy; for example, spot welding, adhesive or sealant application and painting
	2	Applications requiring sensory feedback in order for small modifications to be made to the program to account for variation in the components; for example, arc welding, automotive window glazing and spare wheel mounting
	3	These applications require more complex sensory capabilities such as pattern recognition, that require complex decision making based on this feedback; for example, automated assembly processes such as locating and fixing wheels on a car
	4	The most difficult applications are those involving unpredictable behaviour of either the components or other equipment within the manufacturing cell; for example, operations such as handling of flexible components

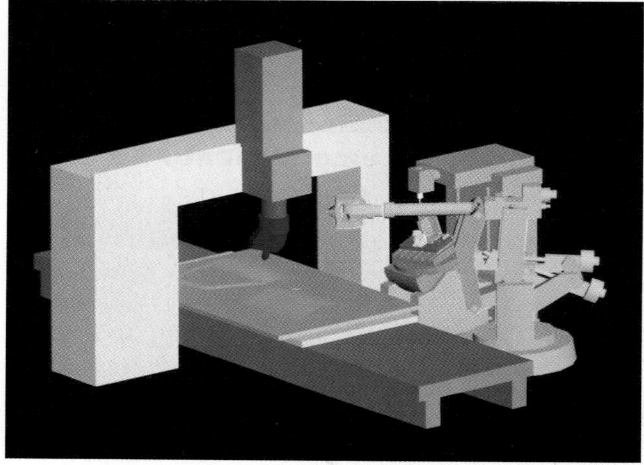

Figure 3.14 and Figure 3.15 *Six degrees of freedom on a robotic arm make it flexible enough to carry out repetitive and precise tasks such as welding*

www.bara.org.uk/info_casestudies.htm

Case studies from The British Automation and Robot Association.

Advantages	Disadvantages
• Ideal for repetitive, monotonous, mundane tasks requiring extreme precision • Can be used in hazardous environments not suitable for human operators • Able to carry extremely heavy loads • Highly flexible when responding to change as they are reprogrammable • Can be programmed once and then repeat the exact same task for years • Do not tire or suffer from lack of concentration and stress during repetitive tasks over long periods • Cost effective as robots can operate continuously resulting in increased productivity • Produce high-repeatability, high-quality products using highly accurate inspection and measurement sensors	• Robots do not have as impressive an array of sensors as humans (touch, vision, hearing, pattern recognition) • Robots do not have the ability to learn and make decisions when the required data does not exist • Robots are not as flexible as humans and are harder to program to perform specific tasks • Robotics technology is extremely expensive to purchase and install in automated manufacturing • Human operators have to be excluded from robot working areas due to safety issues • High cost of making robot cells safe, including collision sensors • Maintenance issues as different brands of robots use different control systems, so maintenance crews need different specialist training • No standard robot programming language implemented, which can cause operating problems between different brands

Table 3.17 *Advantages and disadvantages of robots for manufacturing*

Industrial applications of artificial intelligence

A machine with artificial intelligence (AI) is one that exhibits human intelligence and behaviour and can demonstrate the ability to learn and adapt through experience. The main question is: can a machine really behave like a person?

Research undertaken throughout the 1980s focused upon creating super-computers that could solve problems using reasoning skills like humans. However, humans have a consciousness that gives us feelings and makes us aware of our own existence, and scientists have found it extremely difficult to get robots to carry out simple cognitive (brain/thinking) tasks. Creating a self-aware robot with real feelings is a significant challenge faced by scientists hoping to mimic human intelligence in a machine. For this reason, development since the early 1990s has concentrated on developing expert systems and smaller, autonomous robots that mimic insect behaviour.

Expert systems

Expert systems can apply reasoning skills to reach a conclusion by processing large amounts of known information and coming to a conclusion based on them. Intelligent simulations could generate realistic simulated worlds that would enable extensive testing features for technical design and manufacturing procedures, eliminating the need for physical prototyping. Voice recognition systems coupled with intelligent information resource systems could enable designers and manufacturers to converse with their computers in order to problem solve situations. The system may ask the designer what help they need and automatically call up the appropriate information to help solve the problem. Expert systems could even function as co-workers, assisting and collaborating with design or operations teams for complex systems. This would enhance collaboration by keeping communication flowing among multi-disciplinary teams, program managers and the customer.

Autonomous robots

In manufacturing, the goal is to enable robots to learn the skills needed for any particular environment rather than programming them for a specific repetitive task. Intelligent machine vision systems are key to this as cameras are not as good as human optics. Humans can rely on guesswork and assumptions, whereas robots must comprehend an image by examining individual pixels, processing them and attempting to develop conclusions using expert systems. The advantage a robot may have over humans in this field is the ability to pick up infrared radiation (heat) as well as visible light for thermal imaging or even X-rays.

Autonomous robots in manufacturing must also be able to manipulate and 'feel' objects like humans. At present the typical robot arm allows six degrees of freedom to perform a small range of tasks and can be fitted with complex optical sensors. However, a human type hand is required with over 20 degrees of freedom fitted with hundreds of touch sensors with which the AI system

Figure 3.16 *The future of robotics – Honda's ASIMO can recognise moving objects, human postures, gestures and faces, and environments, and distinguish sounds*

can decide on the best way to manipulate an object from many possibilities. Teams of robots could then contribute to manufacturing by operating in a dynamic environment with minimal human interference.

THINK ABOUT THIS! 

AI sounds like science fiction, which inevitably ends up with machines trying to take over the world, as in *The Terminator*. Should we be worried about creating machines that can think like humans? One of the advantages of humans over robots at present is our ability to think. How will it affect employment, for example, when AI machines become commonplace in manufacturing industries?

Flow charts

A flow chart (see page 109) is a schematic representation of a system, or process, and indicates at which stage QC should take place. It is a visual method of communication rather than text-based, which provides an easy reference for the viewer.

FACTFILE:

Flow chart symbols and their meanings

Symbol	Meaning
	START and **STOP** symbols are represented as lozenges (or rounded rectangles) and indicate the start or end of a process.
	PROCESSES or stages in production are represented as rectangles with the actual process written inside.
	DECISIONS are represented as a diamond (or rhombus) and can indicate where quality control should take place. This symbol has one arrow coming out from underneath for a YES decision and an arrow feeding back into the process stage indicating NO.
	ARROWS indicate the flow of control. An arrow coming from one symbol and ending at another symbol represents that control passes to the symbol the arrow points to.

Open-loop control system

A system operating open-loop control has no feedback information on the quality of each stage of the process. Therefore the process will continue without any interference from the control system even when the output changes. This is a major disadvantage in an automated process, as it cannot detect or correct any errors in the process. An open-loop control system is often used in basic processes because of its simplicity and low cost, especially in systems where feedback is not critical. A light switch is an example of an open-loop system; it is either on or it is off – there is no way of controlling the output.

Closed-loop control system

Systems that utilise feedback are called closed-loop control systems. The feedback is used to make decisions about changes to the process, for example QC. Closed-loop control systems have many advantages over open-loop controllers, including improved tracking of performance throughout a production process and early detection of faults. For the example of the light switch used above, fitting a light-sensor circuit to switch the light on and off at predetermined light levels would give the system feedback, and the light would be switched on and off automatically to suit the user's needs. The system requires no human intervention. Most closed-loop systems are automatic, for example temperature controllers in air-conditioning systems.

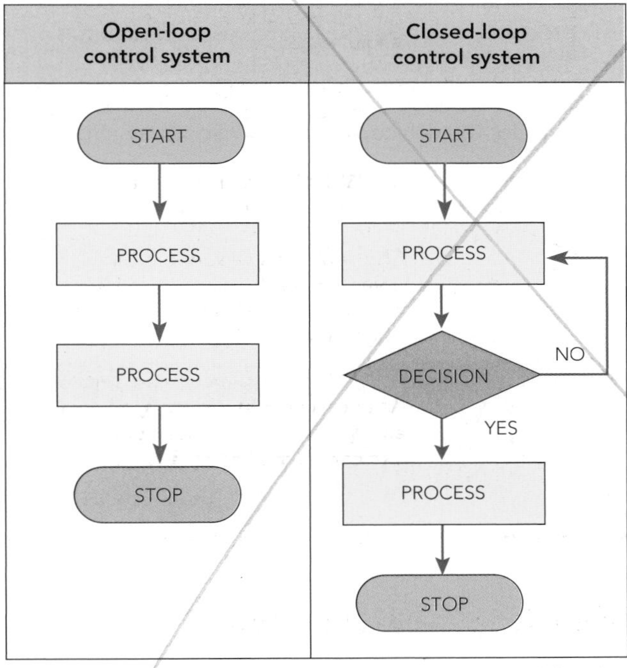

Figure 3.17 *Basic open- and closed-loop control systems*

Text-based solution	Flow chart representation
• Machine is switched on • Chamber needs to fill with plastic granules until it is totally full • Plastic needs to be heated to correct temperature in order to melt • Injection mechanism is operated • Plastic needs to be injected into the mould until completely full • Mould is split and moulded piece ejected • Process is complete	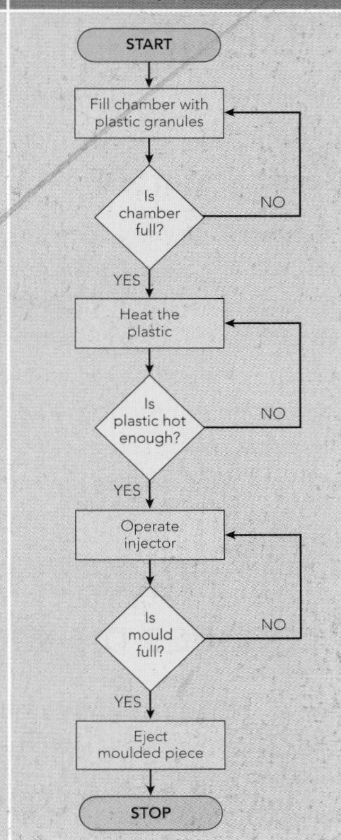

Table 3.18 *Text-based solution represented as a flow chart for the injection moulding process*

THINK ABOUT THIS!

All processes can be visualised as a flow chart for easy communication. Why not try to construct flow charts of simple everyday processes to practise your skills? Chart the stages in making a cup of tea, including areas where decisions have to be made such as amount of sugar. Then move on to more complex systems such as the vacuum forming process that you could undertake in your school or college workshop.

Design in context

Getting started

This section covers a wide range of design-related issues throughout history. Designers must be well informed, not only with current and future applications of technology, but with what has gone before. As a designer you cannot afford to design in a vacuum without external stimulus. Remember: all products were designed by someone for some purpose at some point in time – perhaps they could help with the design of your own product!

The effects of technological changes on society

Design and technology have improved the lives of millions of people worldwide. But the changes brought about by developments in technology have resulted in far-reaching social consequences.

Mass production and the consumer society

The invention of the steam engine by James Watt in 1765 marked the beginning of the 'industrial revolution', which fundamentally changed life worldwide. Industrialisation and specialisation led to changes in production, the workforce, transportation and infrastructure. Many new fields of design were needed to accommodate this and the professional designer came into being. Population explosions occurred in towns and cities where production was centred and a new urban way of life was created. More people needed more products and mass production responded to this need. Expensive and time-consuming crafted work could now be replaced by machine work. Products once exclusively for the rich could now be made at an affordable price for ordinary working people.

Figure 3.18 *1950s youth culture saw the emergence of the 'teenager' as a mass consumer*

The modern mass-consumer society is a feature of the affluent developed world where people's 'wants' are satisfied by a continual stream of new products. It is also referred to as a 'throw-away' culture with an increasing demand for convenience products such as fast food and over-packaged goods.

Mass consumerism, as we know it, developed during the 1930s out of popular culture, lifestyle and fashion. This was a time when international commerce and transportation systems developed and, with them, new opportunities for product design such as luxury ships, aeroplanes, hotels, theatres and department stores designed in an Art Deco style. Innovative new products and materials were introduced, especially in electrical consumer goods such as radios, refrigerators and washing machines. As people's standard of living improved, their demand for new products increased. Advertising and marketing became an important new industry, using market research, packaging and product styling to sell new products. The design of aesthetically pleasing products, most notably 'streamlining', became an important marketing tool.

After the Second World War (1939–1945), for many, there was a period of hardship with very few luxuries as countries struggled to recover from the fighting. By the mid-1950s, however, a new consumer society was developing. It started primarily in America and soon spread to Europe – the 'teenager' was born. Up until this point young men and women wore the same type of clothes as their parents and listened to the same type of music. The advent of Rock 'n' Roll was to change all that and soon teenagers were rebelling against their parents' values and began carving out a style all of their own. Design evolved rapidly to meet the expanding needs of the teenage market, incorporating high fashion and consumer goods such as portable transistor radios for the beach and cars, motorbikes and scooters for the increasingly mobile and independent youth culture.

LINKS TO:

Unit 3.2: Design in context: Influences of design history on the development of products.

Targeting children as new consumers

A rather disturbing feature of modern mass consumerism is the targeting of young children by marketing companies in order to stimulate interest in products at an early age. For example, a new blockbuster movie is released with a wide range of associated merchandise such as toys and lunchboxes that children will pester their parents for. In the past, marketing to children was restricted to toys and sweets and was relatively low-budget. However, as TVs and computers have made their way into children's bedrooms, they can now be targeted not only through TV adverts but the Internet and e-mail as well. It has even been documented that some marketing companies have posed as children in chat rooms to stimulate interest in certain products.

Sue Palmer, in her book *Detoxing Childhood*, outlines the key marketing strategies currently used to target children as new consumers.

- ***'Kids are growing older younger'***, which exploits children's natural yearning to play at being grown-ups and targets them with mini-supermarkets and mini-briefcases that emulate the real thing.
- ***'Gotta catch 'em all'***, which plays on a child's natural urge to collect things, such as Pokemon trading cards, with new sets being released regularly so it is difficult to actually collect them all.
- ***'The culture of cool'***, which plays on a child's need to be accepted by wearing the right brands.

THINK ABOUT THIS! 

What is your opinion on this modern marketing strategy to target young children? Do you think it is right to bombard children with advertising messages at such an early age? What kind of pressure does this put on children and adults?

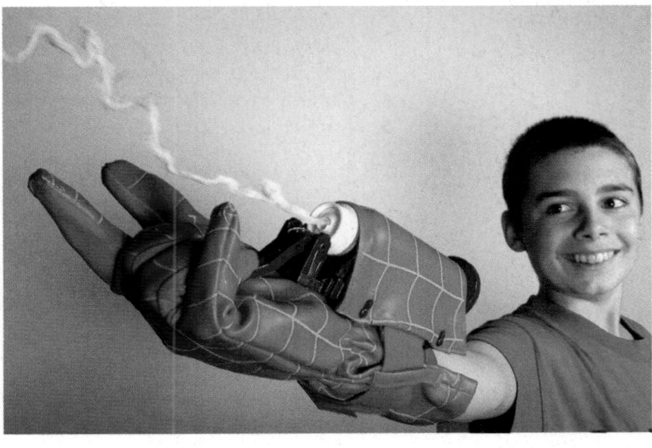

Figure 3.19 *Movie merchandising targets children as new consumers*

Built-in obsolescence

Built-in or planned obsolescence is a method of stimulating consumer demand by designing products that wear out or become outmoded after limited use. In the 1930s an enterprising engineer working for General Electric proposed increasing sales of flashlight lamps by increasing their efficiency and shortening their lifespan. Instead of lasting through three batteries he suggested that each lamp last only as long as one battery. By the 1950s built-in obsolescence had been routinely adopted by a range of industries, most notably in the American motor and domestic appliance sectors. Nowadays products such as laptop computers are obsolete as soon as they are purchased.

Unit 3.4: Sustainability: Cleaner design and technology.

THINK ABOUT THIS!

Make a list, under the headings in the table below, of products that you have consumed over the past 12 months. For example, how many different mobile phones or MP3 players have you owned and where are they now? Does this type of built-in obsolescence bother you?

Mass production and its effect upon employment

Mass production processes, as a result of the industrial revolution, meant that the craftsperson was replaced by low-skilled workers in highly mechanised factories. What started out as a wonderful opportunity for ordinary people to find work and gain access to inexpensive consumer products ended in misery for many. Low skills equalled low wages and the employment of women and children in 'sweatshop' type factories. The resulting poverty led to workers' uprisings and the development of trade unions aimed at combating poor living conditions, poverty and the increasing pollution brought about by industrialisation.

Although working conditions have generally improved, modern mass production still has some very negative social consequences. The use of highly automated production and assembly lines has reduced the workforce required in many factories. The resulting jobs can be divided into two main categories: high-skilled technical roles and low-skilled manual roles. Higher-paid technical roles are required to set up and maintain machinery. Low-skilled and often low-paid workers are utilised on production lines for specialist repetitive tasks, which can lead to very poor job satisfaction and morale.

The 'new' industrial age of high-technology production

In the 20th century, developments in materials and manufacturing technologies, together with changes in lifestyle, revolutionised product design. New materials

Form of obsolescence	Description	Example
Technological	Occurs mainly in the computer and electronics industries where companies are forced to introduce new products with increased technological features as rapidly as possible to stay ahead of the competition.	Mobile phones with image capture almost immediately superseded by phones with moving-image capture.
Postponed	Occurs when companies launch a new product even though they have the technology to realise a better product at the time.	It is not unrealistic to imagine that when Sony launched its PS3 games console it knew what its next generation games console would look like – the PS4?
Physical	Occurs when the very design of a product determines its lifespan.	Disposable or consumable items such as light bulbs and ink cartridges for printers.
Style	Occurs due to changes in fashion and trends where products seem out of date and force the customer to replace them with current 'trendy' goods.	High-street Summer/Winter fashion collections. Premiership clubs update their kit every season so fans need to constantly purchase new replica football kits.

Table 3.19 *Forms of built-in obsolescence*

such as metal alloys, polymers and composites enabled new ways of designing and manufacturing. In particular, the development of digital computers in the 1940s and the silicon chip in the 1960s enabled relatively inexpensive portable computer technology, which transformed modern industrial society.

Computers in the development and manufacture of products

CIM systems incorporating CAD and CAM have revolutionised modern manufacturing and the print industry. The digital age has brought about change to which business has responded by providing quick-turnaround jobs to meet client needs. Printers are capable of producing short full-colour runs with extremely fast delivery times and product designers are able to drastically reduce development times and costs.

On-demand printing quickly supplies the exact amount of copies to satisfy each customer's needs. The use of computers in pre-press means that information can be stored and transferred digitally so designs can be quickly developed in consultation with the client. Once designs are finalised, printing plates can quickly be produced using computer-to-plate (CTP) technology. This cuts out the long process of producing printing plates and instead data is transferred directly to a laser engraver that forms the plate. Printing costs can be significantly reduced with digital printing machines that can operate up to 14,400 pages per hour. Digital printing is well suited to the production of short print runs as it does not require the making of printing plates, unlike commercial printing processes such as offset lithography. In post-press, the printed materials can be die-cut, folded, glued or bound using automated machinery with efficient workflows.

LINKS TO:

Unit 3.1: Industrial and commercial practice: Information and communication technology (ICT).

Unit 3.2: Systems and control: Manufacturing systems, CIM, robotics and artificial intelligence (AI).

Miniaturisation of products and components

The most important technological development in recent years has been in the field of microelectronics. Not only have products reduced in size through technological advances but multi-functional products have become possible. For example, the mobile phone has reduced in size considerably from models first introduced in the 1980s, when most were too large to be carried in a jacket pocket so they were typically installed in vehicles as car phones. The miniaturisation of mobile phones has been possible due to three key developments.

- **Advanced integrated circuits** (ICs) or microprocessors that allow more circuitry to be included on each microchip, increasing functionality and power.
- **Advanced battery technology** including Lithium-Ion rechargeable batteries, providing a lightweight means of storing a lot of energy resulting in smaller and thinner fuel cells.
- **Advanced liquid crystal displays** (LCDs) enabling colour screens that are thinner and brighter and require much smaller current, meaning greater energy efficiency and slimmer housings.

The widespread use of these technologies has also led to advances in manufacturing that have reduced unit costs considerably, enabling low-cost electronic products.

The mobile phone is now much more than a telephone – it has become multi-functional. Communication, entertainment and computing services are converging within the same device, offering substantial choice to consumers. Mobile phones often have features beyond sending text messages and making voice calls. Product convergence has enabled Bluetooth connectivity, Internet access, built-in cameras and camcorders, games and MP3 players to be included on a single device.

Figure 3.20 *The miniaturisation of mobile phones has been possible through advances in technology*

Use of smart materials and products for innovative applications

The continued development of smart materials has seen them being applied to a whole range of innovative products and systems where their ability to respond to changes and return to their original state is a real advantage.

Thermochromic liquid crystals, piezoelectric crystals and smart ink are explored in **Unit 2.1 Materials and components**: New and smart materials

Smart material	Application	Advantages	Disadvantages
Smart glass	Used to change light transmission properties of windows or skylights when a voltage is applied, i.e. changes opacity from transparent to translucent.	• Controls amount of heat passing through a window, saving energy costs. • Provides shade from harmful UV rays. • Provides privacy.	• Expensive to install. • Requires constant supply of electricity. • Speed of control. • Degree of transparency.
Shape memory alloys (SMAs)	Used in spectacle frames as the crystal structure of this advanced composite, once deformed, can regain or 'remember' its original shape, e.g. Memoflex glasses.	• Superelasticity – extremely flexible so can be bent or 'sat on' without permanently deforming. • Immediately recovers original shape. • Lightweight and durable – alloy contains titanium.	• Not unbreakable. • More expensive than similar polymer frames.
Thermochromic pigments	Combined with polymers and used in 'chameleon' kettles, which change colour when boiling (bright pink) and return to original colour when cool (bright blue).	• Immediate visual indication of temperature. • Safety feature. • Aesthetic 'novelty' appeal.	• Limited colour range. • Not possible to engineer accurate temperature settings to colour changes.
Smart fluid/ oils/grease	Used in a car's suspension system to dampen the ride depending upon road conditions, e.g. second-generation Audi TT. The fluid contains metallic elements that alter the viscosity of the fluid when a magnetic field is applied.	• Improves handling and road-holding as it adapts to road. • Better and faster control.	• More expensive than traditional systems.

Table 3.20 *The use of smart materials and products for innovative applications*

Figure 3.21 *Audi's 'magnetic ride' system uses smart fluids to adjust the car's suspension*

The global marketplace

The need to be competitive means that many companies sell their products all over the world. It can sometimes be a problem to design for unfamiliar markets or design products that will sell across different countries. Many companies employ design teams situated throughout the world so they can design for a particular local market or culture. Other companies use focused market research to discover the needs of specific markets.

Offshore manufacturing of multinationals

Offshore manufacture is a driving force in the global marketplace. There is an increased awareness by multinational companies based in developed countries (usually in the West and Australia) of the value of offshore manufacturing as a vital strategic tool. Many companies will draw upon the individual expertise of other countries to develop new products, especially in the field of technology.

Companies are relocating to less-developed countries such as India, China and former Soviet nations and outsourcing their work. Modern corporate buildings and industrial estates are sprouting up in these countries to supply the new demand for outsourcing and offshore manufacturing. Initially jobs in developing countries were created through the manufacture of shoes, cheap electronics and toys, and subsequently simple service work such as processing credit card receipts. Now all kinds of 'knowledge work' and manufacturing can be performed almost anywhere. For example, there is a trend for call centres dealing with the UK public to be based in India.

The driving forces are digitisation, the Internet and high-speed data networks that cover the entire globe. Design data can simply be sent to another country for manufacture or localised expertise can provide the design and development of products. Why do multinationals manufacture offshore, or outsource? The answer is quite simple: it costs them less. It is now possible to receive the same quality work at a fraction of the cost than if Western companies manufactured in their own countries. For example, mould-making for the purpose of injection moulding is generally much more affordable in China than in the West (about 50 per cent lower in China, 30 per cent lower in Taiwan). In addition, by having bases in developing countries it is possible to gain greater access to expanding overseas markets.

Obviously this calls into question certain ethical issues such as large-scale unemployment in developed countries and exploitation of labour in developing countries. For instance, why would a British-based multinational company continue to pay the minimum wage to its UK employees when they could employ Indian or Chinese labour for 50–60 per cent less? However, workers in developing countries may not be given the opportunities for promotion, pay rises, company benefits, union membership and working conditions that their Western colleagues demand as basic human rights. As multinationals build centres of operation and factories in these areas the local workforce is displaced from their traditional trades and become more dependent upon the largely unskilled labour that many industrial processes require.

THINK ABOUT THIS!

Discuss the effects, both positive and negative, of the use of offshore manufacturing and outsourcing in relation to:

a) multinational companies
b) workers in developing countries
c) workers in developed countries.

Local and global production

Issues relating to local and global production are concerned with the effects of the global economy and of multinationals on quality of life, employment and the environment. Whilst the headquarters of multinationals are often located in developed countries, some multinationals are based in developing countries. Though economic regeneration is generally welcomed by the governments of developing countries, there are also a number of negative effects on the local population.

Advantages	Disadvantages
• Economic regeneration of local area through increased employment in manufacturing and service industries. • Improvement in living standards through career development and multi-skilling of workforce. • Physical regeneration of local area through development of infrastructure, transportation and/or local amenities. • Widening of the country's economic base and enabling of foreign currency to be brought into the country, which improves their balance of payments. • Enabling of the transfer of technology that would be impossible without the financial backing of multinationals.	Environmental issues: • increased pollution and waste production as a result of manufacturing activities. • destruction of local environment to build factories, processing plants, infrastructure, etc. Employment issues: • lower wages than workers in developed countries where a minimum wage operates • promotion restrictions as managerial roles occupied by employees from developed countries • no unions for equal rights issues including unfair dismissal/hire and fire • lower safety standards when using 'sweatshops' • devaluing of traditional craft skills, replacement by repetitive 'machine minding' tasks • local community can become dependent on multinationals, leaving community devastated if the multinational pulls production.

***Table 3.21** Advantages and disadvantages of global manufacturing in developing countries*

Influences of design history on the development of products

All design has either been commissioned by or produced for the specific needs of somebody at some point in time. The history of design is inextricably linked to the social, political and economic history of the modern world. Designers have always looked for inspiration from other times and cultures and taken advantage of new technologies, which has had a major effect on the design of their products.

Arts and Crafts (1850–1900)

Philosophy

The Arts and Crafts movement grew out of a concern for the effects of industrialisation upon design, traditional craftsmanship and the lives of ordinary 'working class' people. Although the technical advances of the 19th century brought about new production processes, the design of mass-produced products, such as furniture, was often overlooked. Therefore, poor-quality, over-decorated and often oversized imitations of traditional items of furniture were being produced. This type of furniture was totally inappropriate for the majority of ordinary people

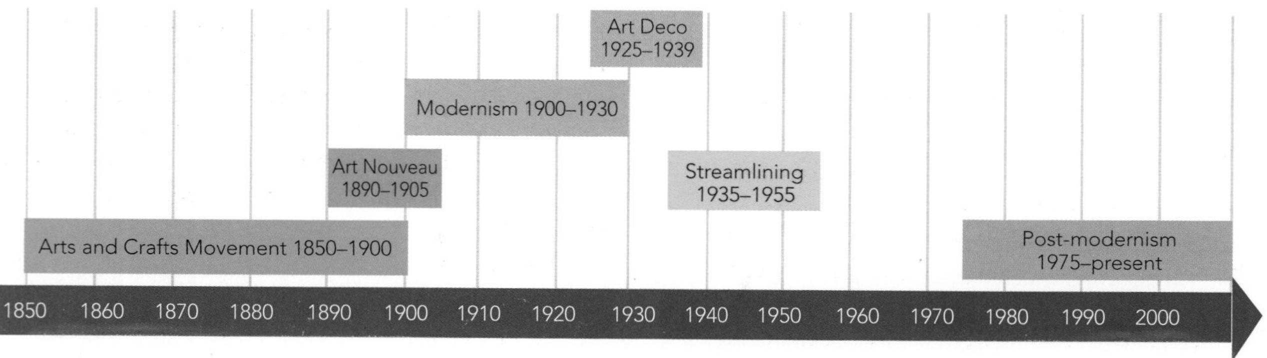

***Figure 3.22** This timeline demonstrates the overlapping of design movements and that one movement did not simply end and another take its place. (Note that the design period between 1955 and 1975 is not covered by the Edexcel specification.)*

who required simple and inexpensive products for their cramped living conditions.

Around this time emerged the two founding figures of Arts and Crafts: John Ruskin (theorist and critic) and William Morris (designer, writer and activist). Ruskin examined the relationship between art, society and labour. Morris put Ruskin's philosophies into practice, placing great value on craftsmanship, simple forms and patterns inspired by nature and the beauty of natural materials. In response to the effects of industrialisation, they helped establish a number of workers' guilds and societies to break down the barriers between architects, artists, designers and makers and pioneered new and unified approaches to design and decorative arts. Their ideas came from the conviction that traditional arts and crafts including weaving, carpentry and stained glass as a 'cottage' industry could change people's lives by empowering the individual as designer/maker of their own products.

Style

- **Simplicity** – Interiors were visually simplistic by removing clutter and including suitably proportioned furniture, which would provide a practical and clean living environment. Furniture was 'humbly' constructed with minimal ornate decoration. The roughness and simplicity of some work was shocking: one reviewer in 1899 referred to an Arts and Crafts piece as looking 'like the work of a savage'.
- **Splendour** – The arts and crafts approach to design led to designers often experimenting with different materials and new techniques in artistic ways. Therefore, small and highly ornate artefacts were produced – working with unusual materials and precious metals.
- **Nature** – Natural plant, bird and animal forms were a powerful source of inspiration. The use of stylised flower patterns emulating the natural rhythms and patterns of plants and flowers were a reflection of a purity of approach. Symbolism with motifs such as the heart symbolising friendship or the sailing ship representing the journey of life into the unknown appear in many pieces of work.
- **Colour and texture** – Colour was used in Arts and Crafts interiors to provide unity and focus. The link between colour and nature was particularly close in Arts and Crafts style. Architects and designers preferred natural materials: stone, wood, wool and linen, and materials that were available locally. Rich materials, highly decorated surfaces and strong colours tended to be concentrated in small areas.

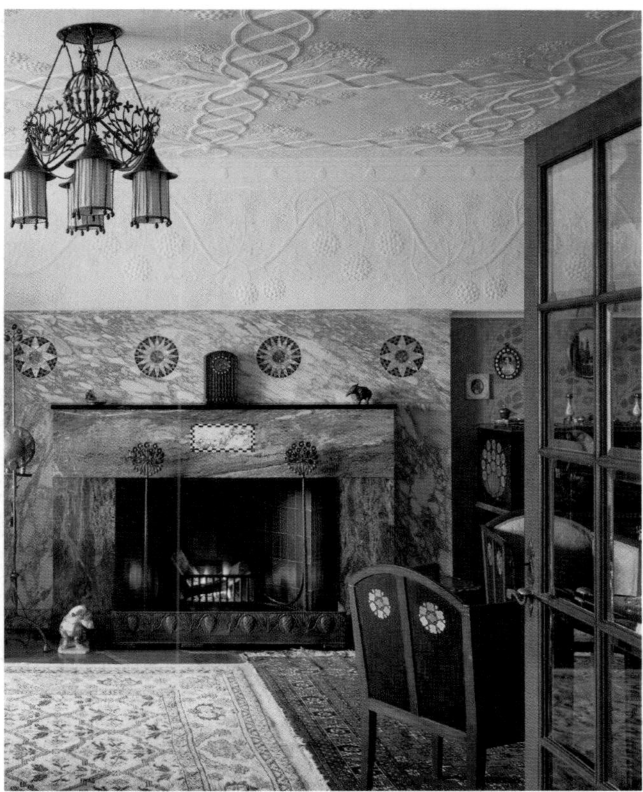

Figure 3.23 *Arts and Crafts interiors were based around the simple rural way of life*

William Morris (1834–1896)

'Have nothing in your houses that you do not know to be useful, or believe to be beautiful.' (William Morris, *The Beauty of Life*, 1880)

William Morris was a poet, writer, designer and innovator in the Arts and Crafts movement but, above all, he was a socialist.

At university, Morris and his friends were influenced by the writings of the art critic, John Ruskin, who praised the art of medieval craftsmen, sculptors and carvers whom he believed were free to express their creative individualism. Ruskin was also very critical of the artists of the 19th century, whom he accused of being 'servants of the industrial age'. After university they formed their own company of designers and decorators with the emphasis placed upon traditional craftsmanship and natural materials. Morris, Marshall, Faulkener & Co specialised in producing stained glass, carvings, furniture, wallpaper, carpets and tapestries. The company's designs brought about a complete revolution in public taste.

Figure 3.24 *The Red House, built in 1859 for William Morris, reflected his 'country cottage' idealism*

Despite the large number of commissions that he received, Morris continued to find time to write poetry and prose and had a number of his works published. His passion for creating 'fantasy worlds' in his novels is said to have had a direct influence on J.R.R. Tolkien's *Lord of the Rings*.

In the 1870s Morris became upset by the aggressive foreign policy of the Conservative Prime Minister, Disraeli, and disillusioned with the subsequent Gladstone Liberal Government. In 1884, Morris co-formed the Socialist League. Strongly influenced by the ideas of Morris, the party published a manifesto where it advocated revolutionary international socialism. Over the next few years Morris wrote socialist pamphlets, sold socialist literature on street corners, went on speaking tours, encouraged and participated in strikes and took part in several political demonstrations (on which he was once arrested). These strong socialist beliefs directly influenced his design philosophy of simple, natural products produced by the individual rather than mass produced by large-scale industry.

Figure 3.25 *The floral ornamentation of Morris's patterns drew heavily upon natural form*

After Morris's death in 1896 the business continued until 1940; a large textile company bought many of Morris's designs and printing blocks for fabric and wallpaper. Many William Morris prints are still in production and have influenced the design style of large companies such as Laura Ashley.

www.artsandcraftsmuseum.org.uk
www.blackwell.org.uk
www.morrissociety.org

Art Nouveau (1890–1905)

Philosophy

Art Nouveau or 'new art' was an international style of decoration and architecture that developed in the late 19th century. The name derives from the *Maison de l'Art Nouveau*, an interior design gallery opened in Paris in 1896, but in fact the movement had different names throughout Europe. It was developed by a new generation of artists and designers who sought to fashion an art form appropriate to their modern age. The underlying principle of Art Nouveau was the concept of a unity and harmony across the various fine arts and crafts media and the formulation of new aesthetic values. It was during this period that modern urban life, as we recognise it today, was also established. Old traditions and artistic styles sat alongside new, combining a wide range of contradictory images and ideas. Many contemporary artists, designers and architects were excited by new technologies and lifestyles, while others retreated into the past, embracing the spirit world, fantasy and myth.

Art Nouveau forms a bridge between the Arts and Crafts and Modernism. There was a strong link between the decorative and the modern that can be seen in the work of individual designers. Many Art Nouveau designers appreciated the benefits of mass production and other technological advances and embraced the aesthetic possibilities of new materials. In architecture, for example, glass and wrought iron were often creatively combined in preference to traditional stone and wood.

Figure 3.26 *Paris Metro station by Hector Guimard, c.1900, creatively combines glass and 'organic' wrought ironwork*

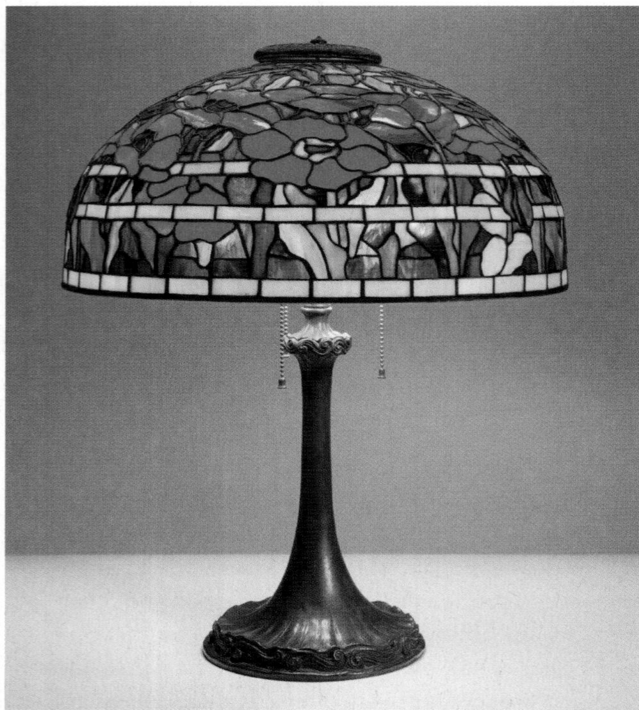

Figure 3.27 *A luxurious Tiffany lamp, c.1900, demonstrates master craftsmanship*

Others deplored the shoddiness of mass-produced machine-made goods and aimed to elevate the decorative arts to the level of fine art by applying the highest standards of craftsmanship and design to everyday objects.

Style

- **Nature** – Like the Arts and Crafts before them, Art Nouveau designers were heavily influenced by natural forms and interpreted these into sinuous, elongated, curvy 'whiplash' lines and stylised flowers, leaves, roots, buds and seedpods. As a complement to plant life, exotic insects and peacock feathers often featured in Art Nouveau designs.
- **The female form** – Art Nouveau is frequently referred to as 'feminine art' due to its frequent use of languid female figures in a pre-Raphaelite pose with long, flowing hair.
- **Other cultures** – The arts and artefacts of Japan were a crucial inspiration for Art Nouveau. Japanese woodcuts, with asymmetrical outlines and the minimal grid structures of Japanese interiors provided vertical lines and height. Celtic, Arabian and ancient Greek patterns provided inspiration for intertwined ribbon patterns.

Figure 3.28 *Mucha's poster for Job cigarette papers, 1898. Interest in Mucha's distinctive style experienced a strong revival in the 1960s and is particularly evident in the psychedelic artwork for many pop groups of this era*

Charles Rennie Mackintosh (1868–1928)

In Britain the Art Nouveau style was exemplified by the work of Charles Rennie Mackintosh. Born in Glasgow, Mackintosh was interested in a career as an architect from an early age, and when he was 16 he became an apprentice to a Glasgow architect, studying at the same time as an evening student at the Glasgow School of Art. It was here that he met like-minded artists and formed the 'Glasgow Four', including his future wife Margaret Macdonald. Through their paintings, graphics, architecture, interior design, furniture, glass and metalwork they created the 'Glasgow style' of Art Nouveau, which influenced many designers throughout Europe.

In 1889, Mackintosh joined the firm of Honeyman & Keppie where he remained until 1913, becoming a partner in 1904.

Figure 3.29 The Glasgow School of Art demonstrates Mackintosh's modern style, which was rooted in traditional Scottish architecture

Figure 3.30 This Mackintosh interior demonstrates his trademark style of a crisp white linear finish dispersed with rose motifs

All his most important architectural and decorative work was achieved during this period. It is clear that he was allowed a degree of autonomy within the firm, developing his own markedly individual style in a way that is not usually possible for a man without his own independent practice. In 1896, Mackintosh won the competition for the building of the new School of Art in Glasgow – a project which gave him an international reputation.

It is perhaps Mackintosh's interior designs that best highlight his goal to create a new artistic harmony. The unification of architectural elements, furniture, furnishings and decoration produced highly aesthetic yet practical domestic and commercial environments. He designed all of the furniture, fixtures and fittings in all of his projects. His style incorporated a contrast between strong right angles and floral-inspired decorative motifs with subtle curves, along with some references to traditional Scottish architecture. His designs for a 'House for an Art Lover' for an international competition in 1901 brought him great praise, although he was disqualified due to late entry. In tribute to his thoroughly modern style the house was built in the 1990s and can be visited by the general public.

Other notable domestic Mackintosh designs include Windyhill, 1900 and The Hill House, 1902. His experimentation with the possibilities of commercial production is best illustrated by The Willow Tea Rooms, 1903.

WEBLINKS:

www.crmsociety.com
www.charlesrenniemac.co.uk
www.houseforanartlover.co.uk

Modernism (1900–1930)

Modernist architects and designers rejected the old style of designing based upon natural form and materials. They believed in 'the machine aesthetic', which celebrated new technology, mechanised industry and modern materials that symbolised the new 21st century. Modernist designers typically rejected decorative motifs in design and the embellishment of surfaces with 'art', preferring to emphasise the materials used and pure geometrical forms.

Modernist principles soon spread throughout Europe with groups including De Stijl in the Netherlands, Bauhaus in Germany, Constructivism in Russia and Futurism in Italy. Le Corbusier, a French architect, thought that buildings should function as 'machines for living in' where architecture should be treated like the mass production of products. This resulted in many high-rise blocks of flats with repetitive 'cubes' as living spaces. Architect Ludwig Mies van der Rohe adopted the motto 'less is more' to describe his minimalist aesthetic of flattening and emphasising the building's frame, eliminating interior walls and adopting open-plan living spaces.

The Bauhaus (1919–1933)

The German economy was in a state of collapse following Germany's defeat in the First World War. A new school of art and design was opened in Weimar to help rebuild the country and form a new social order. Walter Gropius was appointed to head the new institute and named it Bauhaus, meaning 'house for building', which was to combine all the arts in ideal unity.

Philosophy

The central idea behind the teaching at the Bauhaus was a range of productive workshops where students were actively encouraged to be multi-disciplined and trained to work with industry. The Bauhaus contained a carpenters' workshop, a metal workshop, a pottery, and facilities for painting on glass, mural painting, weaving, printing, and wood and stone sculpting. Gropius saw the necessity to develop new teaching methods and was convinced that the base for any art was to be found in handcraft: 'The school will gradually turn into a workshop.' Indeed, artists and craftsmen directed classes and production together at the Bauhaus in Weimar. This was intended to remove any distinction between fine arts and applied arts. The Bauhaus workshops successfully produced prototypes for mass production: from a single lamp to a complete dwelling.

The Bauhaus school disbanded in 1933 when Adolf Hitler and the Nazi party rose to power in Germany before the start of the Second World War. Many Bauhaus leaders, including Gropius, emigrated to the United States to avoid persecution, where they continued to practise. The term 'International Style' was applied to this American form of Bauhaus architecture.

Style

- **'Form follows function'** – Bauhaus featured functional design as opposed to highly decorative design. Designers produced high-end functional products with artistic pretensions which primarily worked well but also looked good. Simple, geometrically pure forms were adopted with clean lines and the elimination of unnecessary clutter.

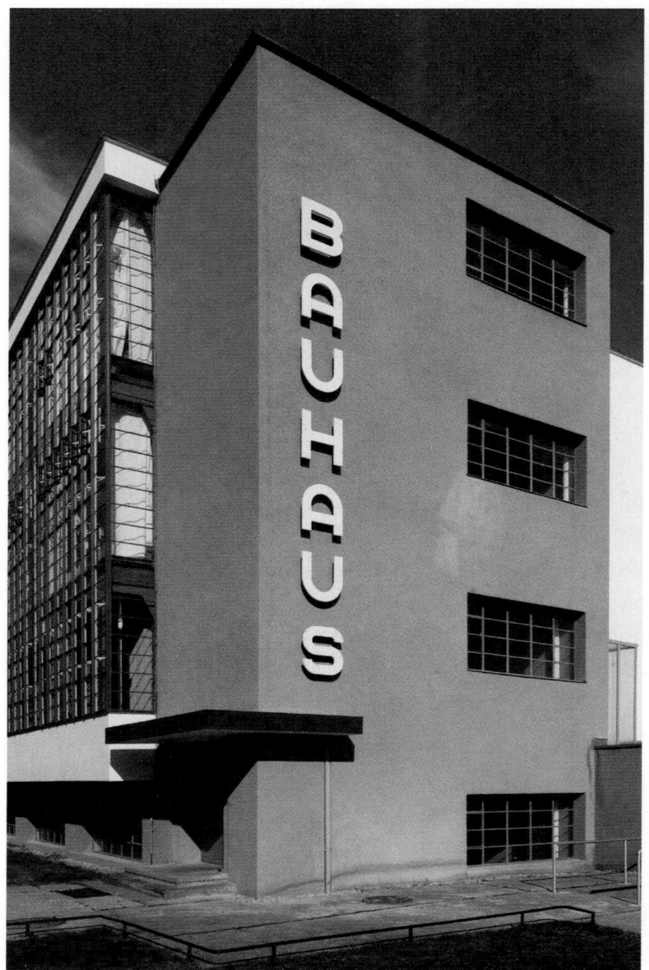

Figure 3.31 *The Bauhaus school, which educated students in art and design with industrial applications*

- **'Products for a machine age'** – Products respected the use of modern materials such as tubular steel and mechanised mass production processes. As a result products looked like they had been made by machines and were not based upon natural forms as with previous movements.
- **'Everyday objects for everyday people'** – Consumer goods should be functional, cheap and easily mass produced so that ordinary working-class people could afford them.

Marcel Breuer (1902–1981)

Marcel Breuer was born in Hungary and worked in an architect's office in Vienna before going to Weimar to study at the Bauhaus from 1920 to 1924. After his trade test he became the manager of the furniture workshop, stressing the combination of art and technology, and created his best-known piece called the 'Wassily' chair. It was here that he met the constructivist artist Wassily Kandinsky who also lectured at the school, but despite popular belief, the chair was not actually designed for Kandinsky.

The Wassily chair, also known as the Model B3 chair, was designed by Breuer in 1925–1926. Kandinsky had simply admired the completed design, and Breuer fabricated a duplicate prototype for Kandinsky's personal quarters. The chair became known as 'Wassily' decades later, when it was re-released by an Italian manufacturer who had learned of the Kandinsky connection in the course of its research on the chair's origins.

Figure 3.32 *Marcel Breuer's iconic 'Wassily' chair demonstrates the principles of Bauhaus design*

The Wassily chair was revolutionary in the use of the materials (bent steel tubes and leather) and methods of manufacturing. The design was only technogically feasible because a German steel manufacturer had recently perfected a process for mass producing seamless steel tubing. Previously, steel tubing had a welded seam that would collapse when the tubing was bent. The Wassily chair, like many other designs of the modernist movement, has been mass produced since the 1960s, and as a design classic is still available today.

In 1937, Breuer emigrated to the United States and received a professorship at the School of Design at Harvard University. In 1946 he founded his own company in New York, Marcel Breuer & Associates, which he managed until his retirement in 1976.

WEBLINKS:

www.bauhaus.de/english
www.artsmia.org/modernism
www.reeform.com

THINK ABOUT THIS!

The principles of modernist architecture were applied to many high-rise blocks of flats built after the Second World War in order to give working-class people modern and affordable housing.

However, a large number of these blocks of flats have since been demolished. Discuss the reasons for failure of many works of modernist architecture with particular reference to the materials used in their construction and the effects of high-rise living upon inhabitants.

Art Deco (1925–1939)

Philosophy

The term Art Deco is widely used to describe the architectural and decorative arts style that emerged in France in the 1920s. It took its name from the 1925 Exposition des Arts Décoratifs held in Paris to celebrate the arrival of a new style in applied arts and architecture. It was an eclectic style that drew on tradition and yet simultaneously celebrated the mechanised, modern world.

Figure 3.33 *This 'Hollywood style' example of Art Deco architecture clearly shows the geometric forms and ocean liner aesthetics of the style*

It embraced both hand-crafted and machine production, exclusive works of high art and mass-produced products in affordable materials.

Art Deco reflected the ever widening needs of the contemporary world. Unlike the stark functionalist principals of modernism, it responded to the human need for pleasure and escape. Art Deco was an opulent style, and its lavishness is attributed to reaction to the forced austerity imposed by the First World War. Geometric forms and patterns, bright colours, sharp edges, and the use of expensive materials such as enamel, ivory, bronze and polished stone are well known characteristics of this style, but the use of other materials such as chrome, coloured glass and Bakelite also enabled Art Deco designs to be made at low cost. This eclectic and elegant style soon became the popular face of modernism and its influence was witnessed worldwide.

Style

- **Geometric forms** – Popular themes in Art Deco were trapezoidal, zig-zagged, geometric fan motifs. Sunburst motifs, for example, were widely used in such varied contexts as ladies' shoes, radiator grilles, the auditorium of the Radio City Music Hall and the spire of the Chrysler Building.
- **Primitive arts** – The simplified sculptural forms of African, Egyptian and Aztec Mexican art and architecture influenced contemporary designers to omit inessential detail. The discovery of Tutankhamun's tomb in 1922 and subsequent exhibition sparked the world's interest in all that was ancient Egyptian and Art Deco responded with some quite literal interpretations.
- **Machine age** – The Art Deco style celebrates the machine age through explicit use of man-made materials (aluminium, glass and stainless steel), symmetry and repetition. Architecture celebrated man's technological achievements in building skyscrapers and ocean liners.

Eileen Gray (1878–1976)

Gray was born in Ireland to a wealthy family of artists and began her university career at the Slade School of Fine Arts in London as a painter. She eventually left painting to study lacquer work under the guidance of Japanese lacquer craftsman, Sugawara. In 1913, she held her first exhibition, showing some decorative panels at the Salon des Artistes Décorateurs in Paris. Here she successfully combined lacquer and rare woods, geometric abstraction and Japanese-inspired motifs into her work.

During the First World War she remained almost permanently in London and only returned to Paris in 1918. Until 1919 she worked as an independent furniture designer, and thereafter as an interior decorator. Her interior designs generated a great deal of praise in the press – amongst her admirers was Walter Gropius, the founder of the Bauhaus. In 1922 she opened the Jean Désert gallery as a showcase for her own designs.

Shortly thereafter, persuaded by Le Corbusier and her lover Badovici, she turned her interests to architecture. In 1924 Gray and Badovici began work on their vacation house, E-1027 in southern France. This is considered to be her first major work, successfully blurring the border between architecture and decoration with a highly personalised design to fit in with the lifestyle of its intended occupants.

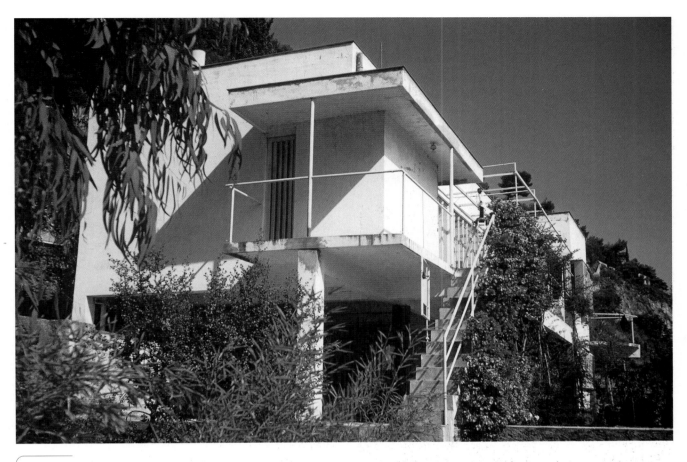

Figure 3.34 Gray's E-1027 villa in France successfully blended her skills as an architect and interior and furniture designer

E-1027 is a codename that stands for the names of the couple: E for Eileen, 10 for Jean (the tenth letter of the alphabet), 2 for Badovici and 7 for Gray. Gray designed the furniture as well as collaborated with Badovici on its structure. Her circular glass E-1027 table and rotund Bibendum armchair were inspired by the recent tubular steel experiments of Marcel Breuer at the Bauhaus. Both pieces of furniture have become design classics and are still produced to this day.

Le Corbusier visited E-1027 on numerous occasions and admired Gray's work very much. Unfortunately for Gray, Le Corbusier loved it so much that he was moved to add his own touch to the clean white villa, painting a series of colouful murals, an act which Gray considered to be vandalism.

www.decopix.com
www.deco-world.co.uk

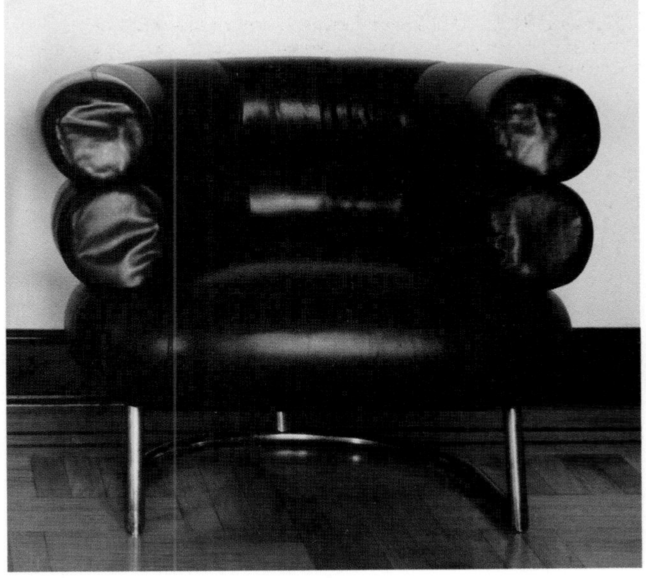

Figure 3.35 *Gray's Bibendum armchair designed for E-1027 is available to buy even today*

Streamlining (1935–1955)

Philosophy

Towards the end of the Art Deco period a new style emerged known as Streamline Moderne, influenced by the modern aerodynamic designs derived from advancing technologies in aviation and high-speed transportation. This was a period of new materials and mass-production processes that could produce more refined products. It was an age when people were looking excitedly to the future and even into outer space.

Streamlining is the shaping of an object, such as an aircraft body or wing, to reduce the amount of drag or resistance to motion through a stream of air. A curved shape allows air to flow smoothly around it. Therefore, in order to produce less resistance, the front of the object should be well rounded and the body should gradually curve back from the midsection to a tapered rear section – a classic teardrop design.

Aerodynamics had been considered by designers for use with automobiles since the turn of the century but it wasn't until the 1930s that new materials and processes were available for cost-effective production. Soon both American and European industrial designers were producing experimental 'teardrop'-based concepts. Although none of these reached production, they had the effect of broadening the minds of the consumer and pointing future design in a new direction.

Figure 3.36 *Norman Bel Geddes patented designs for a teardrop car, bus, yacht, liner and plane although they were never commercially produced*

Figure 3.37 *Perhaps one of the best-known examples of streamlining was the Volkswagen Beetle designed by Ferdinand Porsche. Its origins in streamlining are apparent and it went on to become the most popular and longest-produced automobile ever*

These attractive teardrop shapes were enthusiastically adopted within Art Deco, even applying streamlining techniques to domestic appliances such as radios, vacuum cleaners and refrigerators. Although efficient aerodynamics are not a key feature of many products, the combination of streamline form and modern materials made them stand out from their competitors and therefore more appealing to a growing consumer society.

In the 1950s came the 'Space Age' and the 'Atomic Age' and with it the 'Googie' style of architecture. Googie epitomises the spirit of a generation looking excitedly towards a bright, technological and futuristic age, characterised by space-age designs that depict motion such as boomerangs, flying saucers, atoms and parabolas.

Style

- **Teardrop shape** – With the sleek, efficient forms of airliners and marine life as inspiration, the form adopted as perfect aerodynamicism was that of the teardrop; with the round end being the front. This aerodynamic form became the new aesthetic direction and guided the design of modern products.
- **Futuristic design** – Science fiction provided optimism for a new and better future with sleek rocket shapes and atom designs.

Raymond Loewy (1893–1986)

Loewy was one of the best-known industrial designers of the 20th century. Born in France, he spent most of his professional career in the United States where he influenced countless aspects of American life, from transportation to commercial art.

Loewy launched his career in industrial design in 1929 when a British manufacturer of duplicating machines commissioned him to improve the appearance of one of their products. In three days, the 28-year-old Loewy designed the shell that was to encase these duplicators for the next 40 years. In doing so he was the first designer to transform the look of a product by streamlining, which he referred to as 'beauty through function and simplification'.

Loewy is often referred to the 'father of industrial design' as it was he who first promoted the idea that large corporations could hire industrial designers to provide outside advice on the development of their products. He demonstrated the practical benefits derived from his functional styling, stating:

'Success finally came when we were able to convince some creative men that good appearance was a saleable commodity, that it often cut costs, enhanced a product's prestige, raised corporate profits, benefited the customer and increased employment.'

During the 1930s Loewy established a relationship with the Pennsylvania Railroad, and his most notable designs for the firm were their streamlined passenger locomotives. The GG-1 electric locomotive demonstrated on an even larger scale the efficiency of industrial design. The welded shell of the GG-1 eliminated tens of thousands of rivets, resulting in improved appearance, simplified maintenance and reduced manufacturing costs. As the first welded locomotive ever built, the GG-1 led to the universal adoption of the welding technique in their construction.

During this period Loewy also began a long and productive relationship with US car-maker Studebaker, producing the iconic bullet-nosed cars as well as modernising their logo design. Loewy's car designs incorporated new technological features; introducing slanted windscreens, built-in headlights and wheel covers in his designs, he also advocated lower, leaner and more fuel-efficient cars long before fuel economy became a concern. He was still designing for Studebaker in the 1960s when they launched his most successful car: the 'Avanti' (Italian for forward).

Figure 3.38 *Loewy's aerodynamic streamlining for high-speed locomotives*

As a commercial artist he is credited with designing the classic 'Lucky Strike' cigarette packaging. By changing the package background from green to white, he reduced printing costs by eliminating the need for green dye. In addition, by applying the red Lucky Strike target on both sides of the package, he successfully increased the product visibility and, ultimately, product sales. Other successful projects included the design of the Shell, Exxon and BP logos for the petroleum giants.

From 1967 to 1973 Loewy was commissioned by NASA as a habitability consultant for the Saturn–Apollo and Skylab projects where he helped design the interior living spaces for spaceships. Proven time and again, Loewy's design principles continue to be relevant years later.

www.raymondloewy.com
www.raymondloewy.org

Post-modernism (1975–present)

Philosophy

The term 'post-modernism' was first coined by architect Charles Jencks. He used it to criticise the functionalism of the modernism movement and to describe the eclectic new design styles being developed by a whole range of contemporary architects and designers. The debate regarding whether the term post-modernism, meaning 'after modernism', is appropriate still rages to this day as it does not seem to encompass the range of contemporary thinking and design styles. Indeed, to many the modern movement has not ended as a lot of its ideals are still in use today.

The movement of post-modernism began with architecture, as a reaction against the perceived blandness and hostility present in modernist architecture as preached by the Bauhaus. Its philosophy of an ideal perfection, harmony of form and function and dismissal of decoration was at odds with contemporary designers who wanted individualism and personality back in design.

Figure 3.39 *Richard Rogers' Pompidou Centre in Paris received mixed reviews when it opened in 1978. Its visible support structure and service pipes are a complete contrast to modernist minimalism.*

Out of this period came the the Memphis Group comprising Italian designers and architects who created a series of highly influential products in the 1980s. Founder member Ettore Sottsass disagreed with the approach of the time and challenged the idea that products had to follow conventional shapes, colours, textures and patterns. They drew inspiration from such movements as Art Deco and Pop Art, styles such as the 1950s Kitsch and futuristic themes. Their concepts were in stark contrast to so-called 'good design'. On the launch of the Memphis furniture group in 1981 Sottsass challenged conventional taste by stating:

'Every journalist reacted by saying that the furniture was bad taste. I think it's super taste. It is Buckingham Palace that is bad taste. Memphis relates to the actual world; we are quoting the present, and the future.'

The work of the Memphis Group has been described as vibrant, eccentric and ornamental. It was conceived by the group to be a 'fad' that, like all fashions, would very quickly come to an end. In 1988, Sottsass dismantled the group.

New Design style

- **Humour and personality** – Products were bright and colourful like children's toys, often including unnecessary decoration in an attempt to give static objects personality. By providing products with personality it made them more appealing to the consumer who wanted to express their individuality.
- **'Retro' design** – Designs that take inspiration from past movements and styles and reinterpret them in a modern way. Alternatively, the copying of old designs but manufactured from modern materials and incorporating modern technology to satisfy the trend in nostalgia.
- **Deconstruction** – A development in architecture where the surface structure of a building is distorted so that it becomes non-rectangular. The finished visual abstract and non-symmetric appearance gives the impression of controlled chaos.

Figure 3.40 *Sottsass' Carlton bookcase combines his interest in Indian and Aztec art with 1950s popular culture to produce a bright, colourful and shocking style*

Philippe Starck (1949–)

Philippe Starck is a well-known French designer and probably the best-known designer in the New Design style. His designs have been well publicised in the media and include a diverse range, from spectacular interior designs to mass-produced consumer goods such as toothbrushes, chairs and even houses.

Starck has worked independently as an interior designer and as a product designer since 1975. He rose to fame in 1982 with his interior designs for the French President's private apartments. Since then he has collaborated with many multinational companies on the design of packaging and relatively inexpensive products such a mouse for Microsoft.

Starck's products are often stylised, streamlined and organic in their appearance. They posses humour and he often christens his products with names to bring them to life and give each an individual personality. He values new technologies and has always possessed a taste for innovation with a conviction that 'it's better to make a creative mistake than a stagnant work in good taste'. He is also concious of sustainability and products are often light and economical in their use of energy and materials, from production to consumption, packaging and transportation.

Starck's work for the Italian company Alessi has produced some of the best-known icons of the late 20th century.

LINKS TO:

Unit 3: Design in context: Form and function – another classic Alessi product designed by Starck is the kettle 'Hot Bertaa', which is discussed in the next section.

WEBLINKS:

www.starck.com
www.alessi.com

Figure 3.41 *Starck considers himself as 'a Japanese architect, an American art director, a German industrial designer, a French artistic director, an Italian furniture designer'*

Figure 3.42 *This sleek lemon squeezer named the 'Juicy Salif' was created for Alessi in 1990 and has since become an affordable and popular cult item*

FACTFILE:

Design movements

Movement	Philosophy	Style	Designers	Visual reference
Arts and Crafts	• Fitness for purpose • Honesty in design and making, the return to the designer-craftsman as a reaction against industrialisation	• Simplicity • Natural forms and materials	William Morris Ernest Gimson C.R. Ashbee	
Art Nouveau	• The languid line • The formulation of new aesthetic values for a new urban lifestyle	• Curvy 'whiplash' lines and stylised flowers • Languid feminine form	Charles Rennie Mackintosh L.C. Tiffany Antoni Gaudi	
Bauhaus Modernism	• Functionalism • Reducing form to the most essential elements by omitting decorative frills	• The machine aesthetic using modern materials • Simple, geometrically pure forms and clean lines	Walter Gropius Marcel Breuer Ludwig Mies van der Rohe	
Art Deco	• Popular modernism • Opulent architectural and decorative arts style in reaction to post-war austerity	• Zig-zagged, geometric fan motifs • Symmetry and repetition • Inspiration from ancient Egypt	Eileen Gray Albert Anis Walter Dorwin Teague	
Streamlining	• Consumerism and style • New prosperity and widened consumer choice • Celebrating speed and efficiency	• Aerodynamics • Teardrop shape • Futuristic inspiration	Raymond Loewy Norman Bel Geddes Henry Dreyfuss	
New Design style (post-modernism)	• 'Less is a bore!' expressive and individual as opposed to modernist functionalism	• Humour and personality • Retro design • Deconstruction	Philippe Starck Richard Rogers Ettore Sottsass	

Form and function

The connection between form and function has been one of the most controversial issues in the history of design. When products were first mass produced in Victorian times they were highly decorated to look like hand-made products, whether their decoration was appropriate or not. The development of 'reform' groups such as the Arts and Crafts movement gradually brought about change in the concept of design. The form of products was to be simplified and the products well made from suitable materials. At the turn of the 20th century, developments in materials and technology enabled the production of innovative new products such as the telephone. Many of these products were so innovative that there was no benchmark on which to base their designs.

The development of mass production techniques required that products be standardised, simple and easy to produce. The modernist movement, which supported functionalism, suggested that the form of a product must suit its function and not include any excessive or unnecessary decoration. Therefore, for a product to be mass produced at a profit, it needed to be simple and easy to produce.

FACTFILE:

'Form versus function'

'Form follows function'	Functionality as the prime driver	Functionalists support the view that products should be fit for purpose without any unnecessary decoration.
'Form over function'	Aesthetics as the prime driver	Supporters advocate the aesthetic qualities of products in contributing towards an overall aesthetically pleasing environment.

For many consumers these days, design has become an important means of self-expression. Consumers choose products not just for what they do, but for what they tell the world about their lifestyle choices. Many products are no longer simple, functional artefacts. For example, the purchase of a pair of trainers takes into account many factors such as how comfortable they are, but the overriding reason for buying them may be their appearance and branding. Product performance and reliability are no longer real issues for the consumer as most products carry guarantees and are subject to rigorous quality assurance procedures. The main reason for choosing one product over another with similar functions is its aesthetic qualities.

One of the roles of the designer, then, is to provide the product with the right style or image for a particular target market group. Get this wrong and the product will not sell; get it right then it will become an 'object of desire' for aspiring consumers. Now that so many products are mass produced and sold in their millions, the designer must inject a sense of individuality or personality into an object. For example, the Italian design company Alessi is famous for its playful design of affordable objects and appliances for the kitchen, using bright and colourful polymers and stainless steel to create contemporary and humorous products. However, Alessi did overstep the line between form and function with the Hot Bertaa kettle designed by Philippe Starck, which it had to withdraw from production as it did not boil water very well or safely. This is an important lesson for any modern designer, who must strike a balance between the form and function of the product.

'Form over function'	'Form follows function'
Philippe Starck's Hot Bertaa kettle	
With reference to form	
• Form over function where kettle is a sculptural 'object of desire' and a lifestyle product. • Aesthetically pleasing for fashion conscious, incorporating a brightly coloured polymer handle, aluminium body and cone-shaped shaft that pierces the body of the kettle, serving as both its handle and spout.	• Form follows function where the function of boiling water is the prime driver and the secondary requirement is to look good in the kitchen. • Inoffensive, neutral style that fits in with a wide range of domestic kitchen environments. • Attractive to wide range of customers, with curved handle with ergonomic grip, stainless steel body with contrasting blue water level indicator.
With reference to function	
• Art form where functionality became irrelevant – does not have to look like a kettle in order to boil water. • Poor functionality and user friendliness as the shape proved difficult to pour boiling water from and impractical to fill water through the narrow cylindrical handle. • Intrinsic design flaws and poor safety as heating mechanism largely unreliable, handle became hot once water had boiled, water leaked easily and dangerous to lift as it weighed so much.	• Good functional aspects and user friendliness, including an ergonomically designed handle grip for comfortable pouring, an ON/OFF switch positioned at top for easy access, the handle at the side of the kettle body for easy filling and pouring, a water level indicator that tells the user how much water is in kettle and a large opening lid to fill the kettle. • Important safety features incorporated such as automatic switch off once boiled and the kettle is removable from power supply base so no risk of trailing power cable.

Table 3.22 *Comparison of form and function of two kettles*

THINK ABOUT THIS!

What, in your opinion, is more important – form or function? Can you have one without the other? Or would the ideal product strike a balance between the two? Select a range of similar products, such as trainers or MP3 players, and compare them with reference to both form and function.

Anthropometrics and ergonomics

The relationship between anthropometrics and ergonomics

Ergonomics

Ergonomics is the science of designing products, systems and environments for human use. This means applying the characteristics of human users to the design of a product – in other words, matching the product to the user. Ergonomics is therefore an essential part of the design process. Sometimes products are matched to a single user, where the product is customised to suit one person. The main objective of ergonomics to the designer is to improve people's lives by increasing their comfort and satisfaction when interacting with products, systems and environments. In order to achieve this, data about the size and shape of the human body is required – this branch of ergonomics is called anthropometrics.

Anthropometrics

Anthropometrics is a branch of ergonomics that deals with human measurements, in particular their shapes and sizes. For many products, systems and environments, complex data is required about any number of critical dimensions relating to the user, such as height, width or length of reach when standing or sitting. Therefore, anthropometric data must take into account the greatest possible number of users. This data exists in the form of charts that provide measurements for the 90 per cent of the population that falls between the fifth and 95th percentiles (see Figure 3.43).

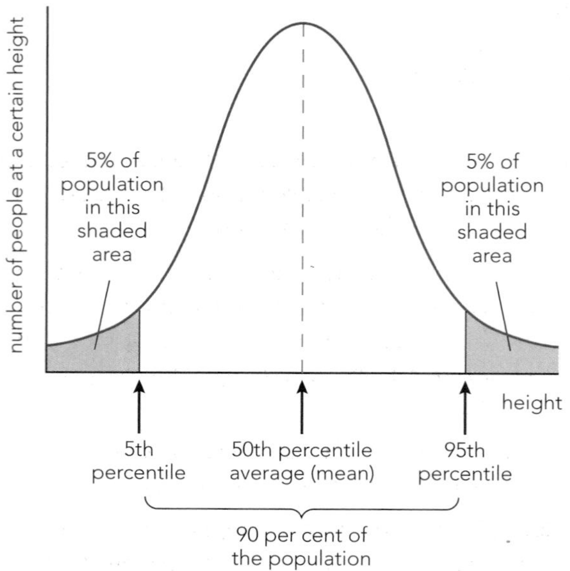

***Figure 3.43** Anthropometric data incorporates human measurements representative of 90 per cent of the population (the fifth to 95th percentiles)*

Sources and applications of anthropometric data

When applying anthropometric data to a design problem, the designer's aim is to provide an acceptable match for the greatest possible number of users. This is achieved by the use of data charts such as those issued by the British Standards Institute (BSI), which are available in a simplified form from the *Compendium of essential design and technology standards for schools and colleges*. Simple data charts relating to measurements for men, women and children can also be found in the clothing sections of mail order catalogues.

Statistical data available from the BSI is associated with heights of men, women and children. The height at which 5 per cent of the population is shorter is known as the fifth percentile. Likewise, the height at which only 5 per cent of people are taller is known as the 95th percentile. The anthropometric data that covers 90 per cent of the British population covers those who fall between the fifth and 95th percentiles. Putting this to use, a furniture designer would have to take into account a range of heights from 1.5 metres to 1.9 metres when designing a chair that would be comfortable for all users. According to the principles of anthropometrics, the designer would ignore the smallest (5 per cent) and tallest (5 per cent) users and design the chair to fit the remaining 90 per cent, who account for the greatest number of users.

Anthropometric data will vary for different regions in the world. For example, the average height of people in Japan is shorter than in the UK. Conversely, the average height of people in Scandinavian countries is taller than in the UK. When designing for the disabled, specialist data is available that takes into account wheelchair use, for example. There are also niches in the market for products and clothing designed specifically for short and tall people with specialist companies and stores.

WEBLINKS:

www.bsi.org.uk/education

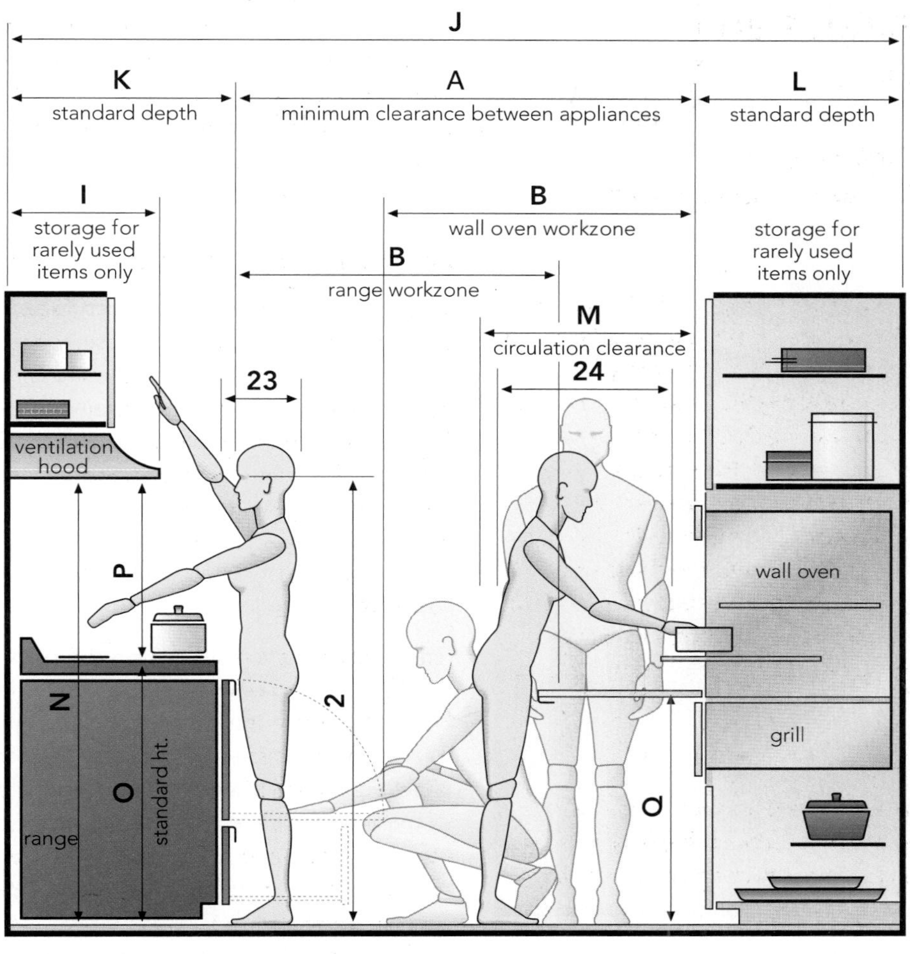

Figure 3.44 *An example of anthropometric data used for kitchen planning*

Key ergonomic factors for the designer

The study of key ergonomic factors by designers is used to improve the design of products, systems and environments by understanding and/or predicting how humans interact with their surroundings. The methods used focus upon different aspects of human performance and take quantitative and qualitative approaches.

- A **quantitative** (measurable) approach predicts the physical fit of the product to the human body shape, encompassing workload, speed of performance and errors.
- A **qualitative** (opinion-based) approach predicts user comfort and their satisfaction with the product, so it has the optimum interaction with the user.

Designers can respect the diversity of human shapes and sizes, making products suitable for both the largest and the smallest people, in four different ways.

- A single design that is valid for everybody: for example, making doorways wide enough for anyone to pass through them, regardless of their body size, the fact that they are carrying something or that they are in a wheelchair.
- Designing a range of objects that covers all possibilities, for example, in clothes sizes.
- Designing a product that is adaptable to different dimensions, for example, a chair whose height can be adjusted.
- Designing an accessory that adapts itself to an original design, for example, car seats for children.

Figure 3.45 *Even ogres need an ergonomic makeover!*

When designing any product it is very important to study the target market group (TMG). The profile of the end-user can often help when designing a product. To this end, ergonomics draws on a wide variety of disciplines, including anatomy, physiology, biomechanics, psychology, design and ICT to make beneficial changes to work and leisure activities to best fit the person carrying out the task. This improves physical and mental health, safety, productivity and efficiency. The following problem was encountered by designers at the Universal theme park in Orlando, USA:

The problem: *Shrek* is a costumed character that performs daily in the theme park. The performers that bring *Shrek* to life have had issues with shoulder, neck and lower-back pains. Due to a number of back, neck and shoulder injuries sustained by the performers who wear the costume, a redesign of the costume was necessary to prevent further injury. The issues identified were weight of the costume, body positioning in the costume, ventilation and visibility. This resulted in 21 first-aid cases, 26 recordable injuries, 622 light duty days and 49 restricted/lost workdays from June 2004 to March 2006 with a direct cost of over US$80,000.

The solution: to achieve the weight reduction, the heavy harness was replaced with lighter materials and additional non-essential layers were removed to reduce it by 10lbs. To straighten body and head positioning, the head was separated from the body to allow movement and to balance weight in line with the spine. A ventilation opening was added into the back of the head to allow for air-flow assisted by a built-in battery-powered fan. The separation of the head, including added mobility to move the neck, resulted in increased visibility. The solution cost just US$100 in materials and US$1,600 for labour and installation – a low-cost, high-yield solution.

The outcome: after changes were implemented on 1 March, 2006 the effects were impressive: injuries fell to zero, with no lost time. The result was increased performance, zero injuries and a simple solution for a low cost of US$1,700, all due to designers listening to the needs of the user instead of making the person adapt to the job.

Adapted from The Ergonomics Society, 15 May 2007

www.ergonomics.org.uk

The interaction between users, products, systems and environments

All types of products, systems and environments are designed so their dimensions suit those of their end-users. Products need to be designed so that they can be operated easily and safely by 90 per cent of the population. Safety considerations require easy operation of products but must also ensure that the product is not operated by mistake.

When designing complex products, systems and environments such as vehicles, designers need to take account of a wide range of ergonomic considerations and anthropometric data covering the different heights and reaches of both male and female drivers.

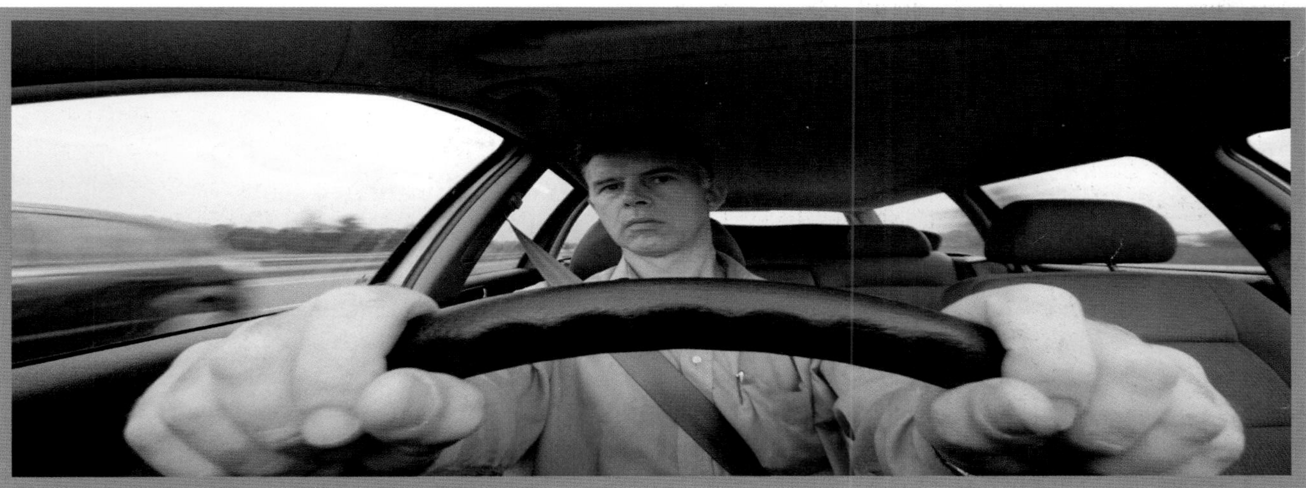

Interface	Ergonomic consideration
Driver's seat	• Seat adjustable backwards and forwards to take into account different leg and arm reaches in relation to steering wheel and foot pedals. • Seat adjustable up and down to take into account different heights so as to see clearly over dashboard. • Shape of seat to take into account different body sizes and padding for comfort. • Seat adjustable for comfort when driving long distances, i.e. lumbar support. • Adjustable head rest to support neck on long journeys.
Steering wheel	• Diameter of steering wheel to aid control of turning. • Thickness of steering wheel, use of soft materials and textures/pistol grip features to provide comfort and safe grip. • Adjustable steering wheel (up and down) to take into account driving styles.
Gear stick/ handbrake	• Positioning in relation to driver's seat. • Diameter of gear knob and handbrake, use of soft materials and textures to provide comfort and safe grip.
Foot pedals	• Distance from seat (adjustable backwards and forwards). • Size and spacing between pedals to take into account different shoe sizes. • Texture on pedal to provide a non-slip surface.
Dashboard instrumentation	• Layout of the dashboard so that all instrumentation is within easy reach of driver. • Essential instrumentation such as speedometer is easily seen (mostly through the steering wheel). • Instant visual impact of instrumentation, i.e. warning/safety lights. • Adjustable brightness of illumination of instrumentation when driving at night. • Size and shape of knobs and switches for ease of use, limiting the need for hand to be removed from steering wheel.
Heating/air conditioning	• Temperature control of the car's interior environment is important for comfort.
Mirrors	• Mirrors that can be fully adjusted from the driver's seat to provide all-round high visibility, i.e. ball joint for manual adjustment of rear view mirror, electric wing mirrors.

Table 3.23 *A driver's interaction with a car interior*

LINKS TO:

Ergonomic considerations when designing computer workstations in **Unit 2.4: Health and safety:** carrying out risk assessments in accordance with the HSE for the design of graphic products using computers and manufacture of models and prototypes using workshop practices.

Sustainability

Getting started

Sustainability means safeguarding the world for ourselves and for future generations, using energy and other resources in a way that minimises their depletion, and designing for a better quality of life. In recent years we have had to rethink our approach to design, materials usage and manufacturing methods by moving towards an approach that considers economic, social and environmental issues and the use of cleaner design and technology. These are not simply issues for governments and large companies – how can *you* contribute towards a more sustainable future?

Life-cycle assessment

It is the case with any design decision and solution that an optimum is looked for and a balance drawn between cost and benefit. Balancing the needs against the impact to the environment is becoming increasingly more difficult for manufacturers as they strive to develop new products and processes. Life-cycle assessment (LCA) is a technique now widely used to assess and evaluate the impact of the product or packaging 'from the cradle to the grave' through the extraction and processing of raw materials, the production phase, and life-cycle processes including distribution, use and final disposal of the product.

Life-cycle inventory

Consumers are becoming increasingly aware of environmental issues and expect companies to pay attention to the environmental impact of their products. However, British Standards and the ISO 14000 series of standards now demand continuous improvement in a company's environmental management systems, of which a life-cycle inventory is an important aspect.

A life-cycle inventory describes which raw materials are used and what emissions will occur during the life of a product. The basis of this study is to collate an objective inventory of all the inputs and outputs of industrial processes that occur during the life cycle of a product, including:

- **environmental inputs and outputs** of raw materials and energy resources
- **economic inputs and outputs** of products, components or energy that are outputs from other processes.

The life-cycle inventory can be expressed as a process tree (see Figure 3.47 below) where each box represents a process with defined inputs and outputs that forms part of the life cycle. The second stage of this process is to interpret the data to assess the overall environmental impact of the product.

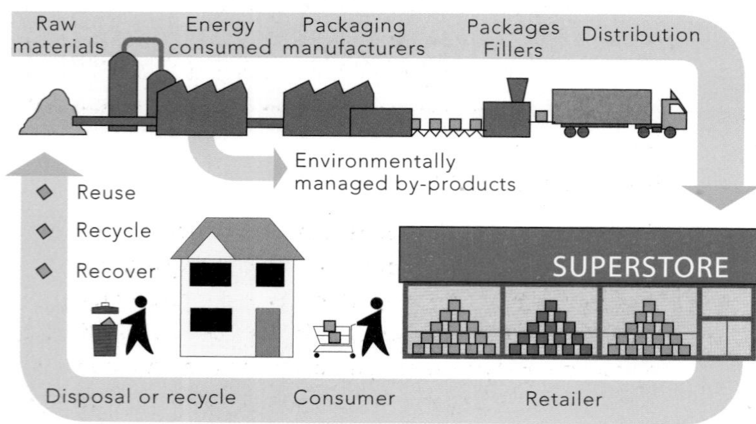

Figure 3.46 *Life-cycle assessment (LCA) for packaged products*

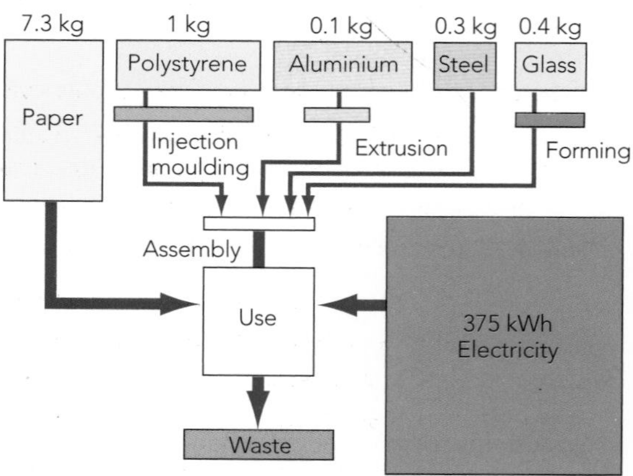

Figure 3.47 *This simple process tree of an LCA inventory for a coffee machine clearly indicates design priorities: minimise the use of electricity and paper filters*

Cleaner design and technology

Sustainable product design

Successful cleaner design requires reducing the environmental impact of the product or packaging throughout its entire life cycle. Issues such as raw material use, waste production, energy consumption and emissions into the atmosphere all need to be considered at each stage of the product life cycle. It is now fundamentally important for a designer to also consider *design for recycling*, including the following requirements of a product:

- easy to dismantle for repair or reuse and so extending product life
- easy to separate different materials for recycling
- easy to remove components that must be treated separately for repair
- use as few different materials as possible
- mark the materials/polymers in order to sort them correctly
- avoid surface treatments in order to keep the materials 'clean'.

In his book *The Total Beauty of Sustainable Products* (published by RotoVision, ©2001), Edwin Datschefski spells out five factors that make up product sustainability.

1. **Cyclic:** products that are made from biodegradable organic materials or from minerals that are continuously recycled, decreasing levels of waste and pollution; for example, products made from Biopol®.
2. **Solar:** products that in manufacture and use consume only renewable energy that is cyclic and safe; for example, products made using renewable energy sources such as wind power and products that operate using solar-powered (photovoltaic) cells.
3. **Safe:** all releases to air, water, land or space are 'food' for other systems; for example, products that do not emit unnecessary pollutants or chemicals during their manufacture.
4. **Efficient:** products that in manufacture and use require 90 per cent less energy, materials and water than equivalent products did in 1990; for example, a reduction of materials used in packaging a product, which decreases the amount of raw materials extracted, energy used in processing/manufacturing, and pollution, etc.
5. **Social:** products whose manufacture and use supports basic human rights and natural justice; for example, Fairtrade products that help producers in developing countries receive a fair share of profits, which reduces exploitation of the workforce.

FACTFILE:

Key environmental considerations for cleaner design and technology

Life-cycle stage	Key environmental considerations
Raw materials	• Use less material • Use materials with less environmental impact • Consider recyclable materials • Adhere to relevant legislation
Manufacture	• Reduce energy use • Simplify processes where appropriate • Reduce waste • Use natural resources efficiently
Distribution	• Reduce or lighten packaging • Reduce mileage of transportation to customer
Use	• Increase durability of product • Encourage refill consumables where appropriate • Use 'green' credentials as a positive marketing strategy • Promote efficient use of product
End-of-life	• Make reuse and recycling easier • Reduce waste to landfill

Raw materials

The key issues for designers when considering the use of materials for products and packaging are the environmental and economic costs of the raw material. Although metals are abundant in the Earth's crust, their extraction and processing is costly both environmentally and financially, largely due to the vast amounts of energy required to convert the ore into a finished product (for example, a drinks can).

Usable material	Raw material	Extraction	Processing
Timber	Trees	Deforestation, environmental degradation of forest areas, distribution, etc.	Chemical pollutants used in chemical wood pulp production and bleaching.
Metals	Aluminium – bauxite ore Steel – iron ore	Environmental impact of mining activities, e.g. energy use, open-cast mining, transportation, etc.	Vast amounts of energy required to process ores and resultant carbon dioxide emissions.
Polymers	Crude oil	Environmental impact of drilling activities, e.g. energy use, destruction of habitat, etc.	Vast amounts of energy required to refine oil and produce polymers, with resultant carbon dioxide emissions.

Table 3.24 *Environmental impact of raw materials used in manufacturing processes*

Polymers are derived from crude oil, which is a finite resource and for this reason designers must consider the use of recycled materials to reduce consumption. The UK relies heavily upon imported raw materials which have to be transported long distances, resulting in high transport costs and carbon dioxide emissions.

The answer, then, is relatively simple: reduce the amount of materials used in order to conserve resources, which will in turn reduce energy consumption and pollution, and use more recycled materials or use materials that are recyclable. Of course, with current mass production and mass consumerism the solution is not an easy one.

WEBLINKS:

www.globaltrees.org/proj.asp?id=40

This site has some interesting information regarding the management of timber and some good links to specific woods.

Manufacture

The conversion of raw materials into finished products incurs considerable environmental impact and costs. For many companies the analysis of existing manufacturing processes can identify areas that can be modified to achieve more efficient and cleaner processes. The aim is to reduce production costs by creating designs that use less material and less energy during manufacture and to reduce waste production. Modifying the design and manufacture to increase efficiency may involve using:

- a simpler design with fewer components to reduce materials use and assembly time
- different materials to reduce their weight or the quantity used
- materials that use less energy during manufacture and produce less waste
- simpler components that are easier to machine or mould and produce less waste
- a simplified or different work flow with improved quality control.

The Coca-Cola '202' drinks can

Coca-Cola Enterprises Ltd produces and distributes around two billion soft drinks cans per year. In the early 1990s, many people switched to plastic bottles, putting pressure on can manufacturers to reduce costs. As further lightweighting was not thought possible with the existing '206' can design, a new '202' can was developed. It had a reduced end diameter, whilst maintaining the same body diameter to contain the same volume of liquid.

The change in design had to take account of the can manufacture, the filling process, the can strength and stackability for distribution. Coca-Cola secured the agreement of European can manufacturers to a common specification for the '202' can end and body. Previous to this agreement machinery had to be reset to accommodate different suppliers' cans and different end profiles.

The new '202' can design successfully reduced raw materials costs and simplified the production processes, resulting in:

- cost savings of over £1 per thousand cans and reduced metal use worldwide in the canned drinks industry

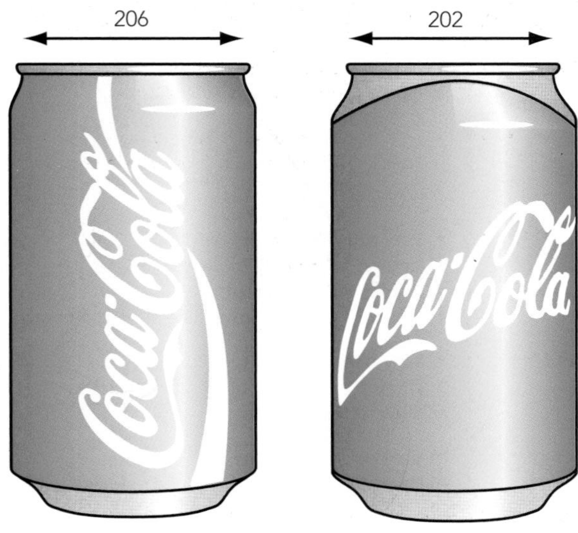

Figure 3.48 The Coca-Cola '202' can is a good example of product lightweighting

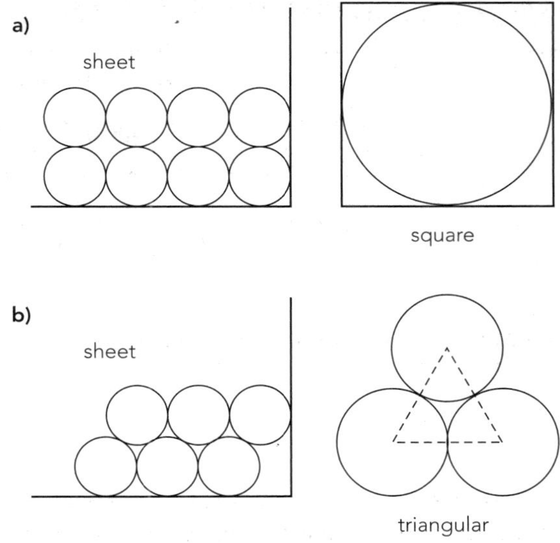

Figure 3.49 Efficient lay planning reduces material wastage

- savings of around £2.3 milllion a year from 1995 onwards for Coca-Cola
- the use of lightweight cans enabling more products to fit in a single lorry, reducing the number of journeys in distribution.

The possibility of reducing the amount of materials used to manufacture a product is often found in processes that involve cutting and stamping shapes from sheet materials. For example, in can manufacture careful calculations and efficient lay planning must be made to limit the amount of aluminium used for making the circular tops of the cans. There are two methods of arranging the can top on a rectangular sheet: with the circular tops in either a square or triangular formation. When the wastage of materials from the two methods is calculated it works out that the square method produces 21.4 per cent scrap, with the triangular method producing just 9.3 per cent scrap. This scrap material can be recovered and recycled. Clearly efficient lay planning has an enormous impact upon reducing material wastage.

Distribution

There are a number of issues relating to cleaner distribution of goods around the UK but they all result in the same key concerns – extremely large energy use and resultant carbon dioxide emissions, which contribute towards global warming. Congestion on our roads and motorways is increasing and road haulage companies are significantly adding to this. Other forms of transport could be used that are less polluting, such as trains (especially electric trains) or even waterways where appropriate. The size of journeys could be reduced from the manufacturer to the consumer either by use of local resources or geographical locations of distribution centres. If a lorry has to make a journey then there are a number of things that could be done to save fuel, such as reducing or lightening the amount of packaging used in products, driving sensibly and smoothly and exploring alternatives to fossil fuels.

Alternatives to fossil fuels

Apart from the obvious pollution from traditional fossil fuels, the financial cost of diesel and petrol will continue to rise. Therefore, the only realistic course of action for drivers and transport companies is to find less polluting and cheaper alternatives that address three key factors: good performance, reliability and availability. Unfortunately, it is the lack of availability of these alternative fuels that is the main reason they are not widely used at present.

Fuel type	Advantages	Disadvantages
Liquefied petroleum gas (LPG)	• Relatively good fuel availability • Good range of kits available • Reduced emissions • Increasingly good supply of used vehicles • Low-cost fuel – less than 50 per cent of diesel • Reliable performance	• Not available for diesel vehicles • No factory-fit models available
Bio ethanol	• Reduced emissions • Increased power • Factory-fit models now available • Renewable fuel	• Very poor availability of fuel • Limited availability of vehicles • Similar price to diesel • Up to 30 per cent lower economy than petrol
Compressed natural gas	• Kits fit to existing diesel vehicles, such as HGVs • Similar economy to diesel • Reduces emissions	• Very poor availability of fuel • Limited availability of kits and vehicles • Slow refuelling times
Hydrogen	• Zero emissions • Renewable fuel	• Very poor availability of fuel • Limited availability of kits and vehicles
Electricity	• Zero emissions	• Very limited range • Slow charging/refilling time

Table 3.25 *Advantages and disadvantages of alternative fuels*

Figure 3.50 *The Toyota Prius was one of the first mass-produced hybrid electric vehicles that switches between an electric motor for city driving and an efficient petrol engine for the open road*

www.dft.gov.uk/ActOnCO2

Department for Transport website devoted to cutting carbon dioxide emissions.

www.toyota.com/prius

Information on the Toyota Prius hybrid electric vehicle.

Use and maintenance

The designer's responsibilities do not end after the product reaches the consumer – reducing the impact from the use of the product must also be considered. Many products are designed and manufactured in such a way that it makes it virtually impossible to access internal components if something stops functioning.

This 'built-in obsolescence' means that the product cannot be repaired and therefore has to be discarded and replaced. The issue here, then, is one of repair versus replacement. An example is the Sennheiser DJ headphones which have replaceable parts for extended product life. The parts that are most likely to become damaged, such as tearing the ear pads or pulling out the cable, can be ordered individually from the company and are easily replaced.

Figure 3.51 *These Sennheiser HD 200-V1 DJ headphones have replaceable parts for extended product life*

There is also the issue of upgrading technology when it becomes obsolete. Mobile phones have developed considerably over the last few years and each new function means that the old phone becomes redundant. Perhaps there is a case for companies to simply offer an upgrade service where the handset can be retained. If upgrades are not available then the phone could be reused in a secondary market, e.g. developing countries where communication in remote areas is the priority and multi-functions such as games are not an issue.

THINK ABOUT THIS!

Have you ever purchased a product and asked whether parts could be replaced if they fail? Why do you suppose that this is not a major issue for many people and how could the repair and not replacement of products be encouraged? Have manufacturers got to encourage this or should the attitudes of society change?

Design in context: The effects of technological changes on society: built-in obsolescence.

Minimising waste production

Perhaps the most important economic factor for a designer of sustainable products to consider is that waste is lost profit. There are some simple options to consider when deciding how to minimise waste production at the end-of-life stage; they are referred to as the four Rs:

- reduce
- reuse
- recover
- recycle.

Reduce

For all designers, one of the first priorities for sustainability should be to reduce the quantities of any material chosen whenever possible. Therefore, packaging designers must optimise the amount of materials needed to package a product in order to minimise the consumption of resources, which will in turn achieve significant cost savings and improve profit margins.

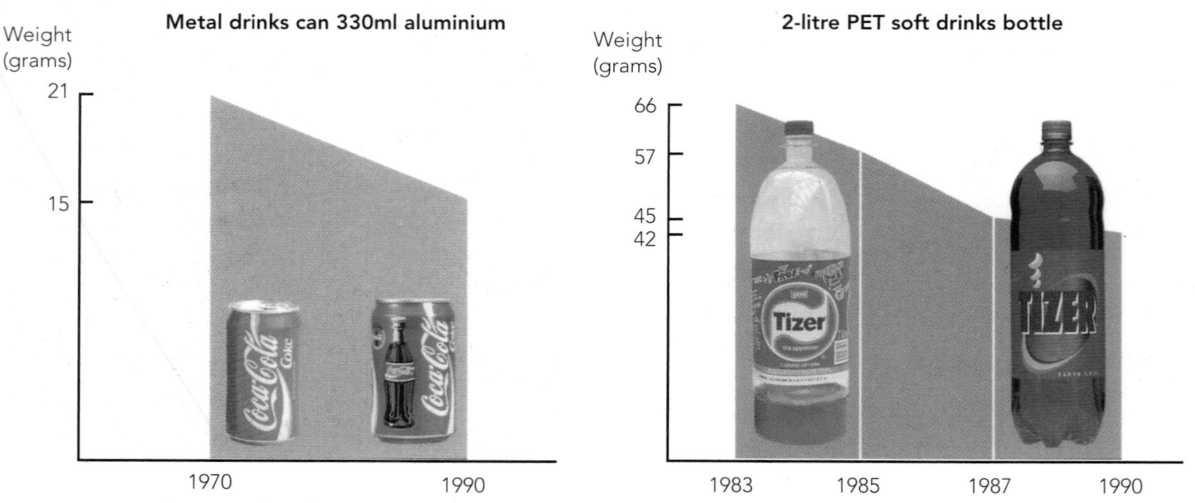

Source: The Industry Council for Packaging and the Environment (INCPEN)

***Figure 3.52** All types of packaging materials have evolved to contain the same volume of goods with less weight of material*

Manufacturers are obliged to reduce packaging use under the UK Producer Responsibility Obligations (Packaging Waste) Regulations (1997). The Government's Envirowise programme suggests that manufacturers:

- consider the materials and designs they use
- examine ways of eliminating or reducing the packaging requirement of a product – changes in product design, improved cleanliness, better handling, JIT delivery, bulk delivery, etc.
- optimise packaging use, i.e. match packaging to the level of protection needed.

In one Envirowise case study, Ambler of Ballyclare implemented an environmental policy focusing upon the minimisation of packaging waste. The financial benefits were considerable (£103,002 savings per year), as were the environmental benefits including diverting waste from landfill,

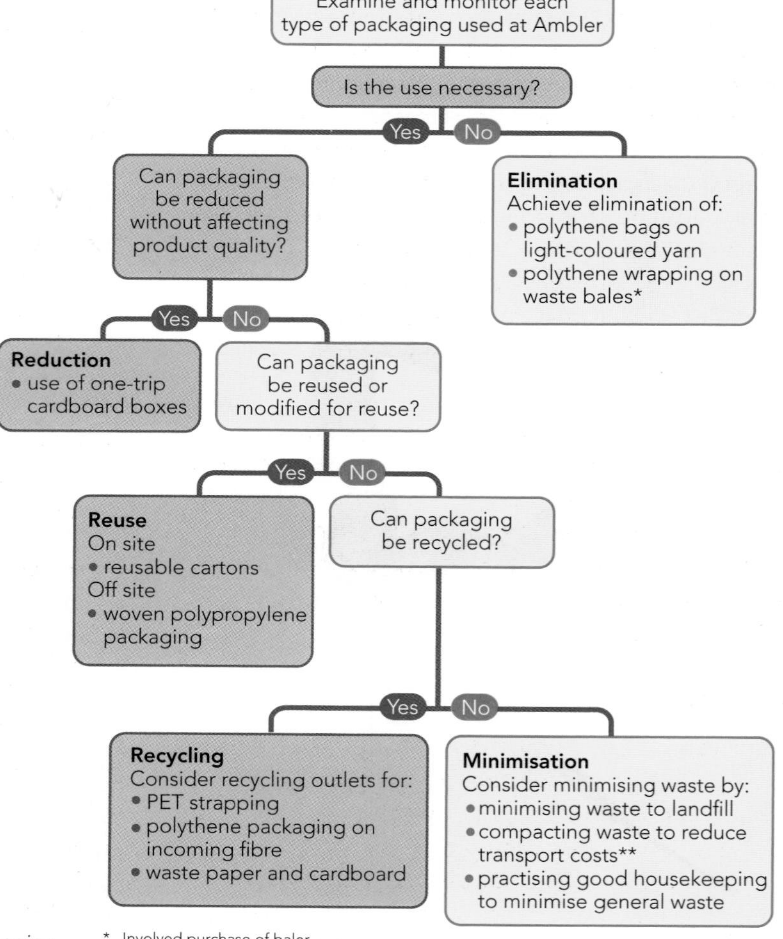

* Involved purchase of baler

** Involved purchase of compactor

***Figure 3.53** Systematic approach to identifying packaging use and waste minimisation by a company*

reducing environmental impact and reducing disposal costs. The reduced use of vehicles for the transportation of waste also saw a 42 per cent reduction in carbon monoxide emissions.

www.envirowise.gov.uk

Independent advice and support on practical ways to increase profits, minimise waste and reduce environmental impact for UK businesses.

Reuse

Part of the cyclic factor of sustainable design is the reuse of products, which minimises the extraction and processing of raw materials and the energy and resources required for recycling.

A number of companies adopt returnable, or refillable, containers for some of their products, for example the door-step delivery of milk in glass bottles. Refillables appear to offer environmental benefits yet they often require greater use of resources in their manufacture and distribution to enable them to withstand the rigours of repeated use. This initial use of extra resources can be offset by the reuse of the container but only in local distribution and collection schemes. If reuse is to be economically viable then the cost of collection, washing and refilling should be less than producing a new container.

Refillable containers have been one of the most dramatic developments in retailing in recent years, most notably in the reduction of packaging size of detergents and fabric conditioners. Concentrated forms of these products have been realised through technological developments, which has resulted in less packaging per dose of detergent.

Recover

The manufacture of any product obviously requires the use of energy. If the product is simply discarded and landfilled then all of this energy is lost. Waste that cannot readily be recycled but can burn cleanly can be incinerated in specialised power stations to generate electricity and provide hot water for the local area. This is not an ideal solution (waste reduction is), but by adopting such technology less finite fossil fuel is needed to generate electricity in conventional power stations. In Sweden 47 per cent of waste is recovered in energy from waste plants and in the Netherlands 34 per cent is recovered. Tetra Pak education service has claimed that you could run a 40W bulb for an hour and a half on the energy released when one aseptic carton is burnt.

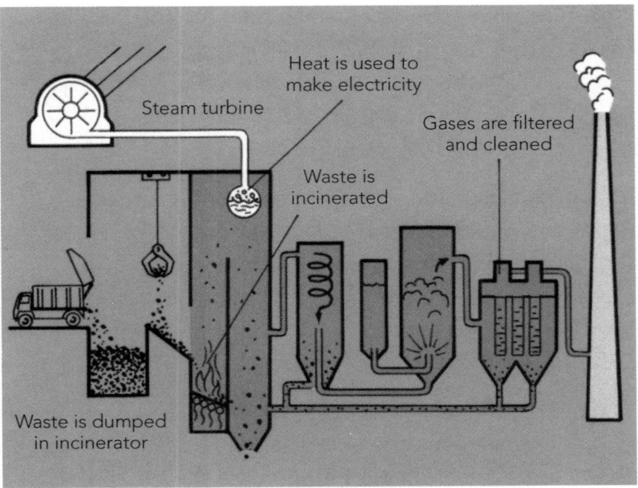

Figure 3.54 *Recovering energy from waste*

Recycle

Essentially, recycling takes waste materials and products and reprocesses them to manufacture something new. Some materials, such as paper and boards, can be made into the same products and others can be made into something completely different such as plastic vending cups into pencils. Recycling is an important aspect of a modern consumer society with millions of tonnes of waste being disposed of in landfill sites or incinerated, causing environmental concerns.

Recovery and recycling of metal

Metals are ideally suited to recycling as they can be readily melted down and reused many times. The first stage in the recycling of metals is that the ferrous and non-ferrous metals need to be separated out.

Ferrous metals such as steels and cast irons are graded by size prior to being melted down. They have a relatively low scrap value and this can be an issue as it could be more economical to use steel produced from raw materials as opposed to using steel manufactured from scrap. Although there is a case against the recycling of steel on economic grounds it is the world's most recycled material.

Non-ferrous metals, such as aluminium and copper on the other hand, are of more value in their scrap form. These metals are sorted into different grades of material and are used according to their particular grade.

One of the issues of recycling metal is that in some cases it is very difficult to tell the difference between ferrous and non-ferrous scrap. A drinks can may be manufactured from either steel or aluminium and when they are covered in graphics it is difficult to tell what the material is.

The best test in this instance is to check the can with a magnet. Steel is magnetic and aluminium is not. This characteristic is often exploited in the recycling plant. As the scrap material passes through the plant on a conveyor belt it moves beneath a large electromagnet that separates out ferrous from non-ferrous metals. Once this is done, the steel is sent off to be melted down in one furnace, which is usually an electric arc furnace, and the aluminium sent to a second furnace. One interesting fact in recycling aluminium is that it actually takes less energy to recycle aluminium than it does to produce aluminium from bauxite.

Metal recycling	
Steel	Using an electric arc furnace scrap metal is processed into high-quality tool steel and stainless steel.
Copper	75–80 per cent scrap copper is processed in blast furnaces, or electric arc furnaces, to produce high-quality copper.
Aluminium	To produce aluminium from scrap requires temperatures of about 660°C. To produce 'virgin' aluminium using the Hall-Heroult process requires temperatures of about 900°C.

Table 3.26 *Metal recycling*

WEBLINKS:

www.recyclemetals.org/whatis.php

www.recyclingexpert.co.uk/RecyclingMetals.html

These websites give an insight into the recycling of metals.

Remarkable pencils

Remarkable Pencils Ltd has successfully managed to recycle plastic vending cups into pencils and other stationery. The plastic contained in one high-impact polystyrene vending cup is enough to make one remarkable pencil. Some people may argue that plastic vending cups are themselves unnecessary but daily production rates of 20,000 pencils means that 20,000 vending cups are being saved from disposal on landfill sites every day. The company has developed a strong, contemporary brand identity with its products so that consumers are more liable to purchase an environmentally aware alternative to the traditional wooden pencil.

THINK ABOUT THIS! 

Conduct a study on the amount of waste generated by your school or college. Where does the majority of it come from – photocopying, litter, school dinners? How could this waste be minimised? Just think that your school is just one of thousands of schools in the UK alone – all producing as much waste as you!

Renewable and non-renewable sources of energy

Since the industrial revolution in the 18th century, the burning of fossil fuels on an increasingly massive scale for generating power has resulted in large emissions of carbon dioxide, a greenhouse gas that contributes to global warming. The concentration of carbon dioxide in the Earth's atmosphere is increasing, raising concerns that solar heat will be trapped and the average surface temperature of the Earth will rise. Scientists estimate that a rise in global temperatures of between 1.5 and 2°C could cause unimaginable devastation through flooding, changes in weather patterns, desertification and displacement of entire populations. Therefore, a more sustainable solution for the planet's future energy needs based upon economic and environmental implications must be sought.

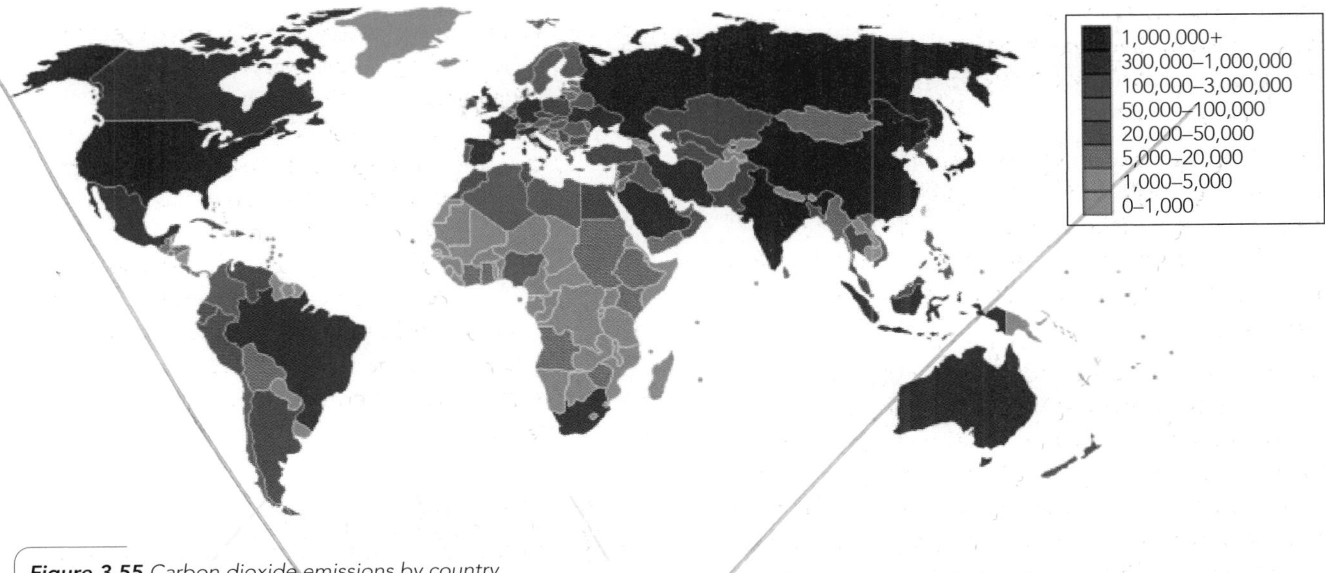

Figure 3.55 *Carbon dioxide emissions by country*

The 'Sandals' versus the 'Nukes'

Since global warming has become high on the environmentalist agenda, 'green' groups are split on the issue of which is the more sustainable – renewable or non-renewable forms of energy.

On one side is the conventional green lobby, which believes in weaving ourselves more deeply into the natural world through the use of renewable energy sources like wind, solar and biofuels. On the other side is the less conventional opinion that humans should not place more demands upon nature but instead use technology such as nuclear energy and carbon-scrubbed natural gas.

The ideal power source has to produce the largest amount of energy achievable at an affordable cost, with as little environmental pollution as possible. Renewable energy is still in its infancy and is currently quite expensive to set up, but with time and more widespread use could become economically viable. However, its supply is unreliable (sun for solar cells and wind for wind turbines) and there are problems with energy storage. Nuclear energy, on the other hand, is now an extremely safe way of producing energy with very low carbon emissions. Many scientists argue that a combination of the two is the best way forward.

FACTFILE:

'Sandals' versus 'Nukes'

Sandals	The Sandals believe in the back-to-nature approach where humanity should be woven more deeply into the natural world through the use of renewable energy sources.
Nukes	The Nukes are less conventional green thinkers who believe that technology is the way forward, therefore reducing our dependency on the planet's natural resources.

THINK ABOUT THIS!

Divide your class up into two groups: one will be the Sandals and the other Nukes. Research and hold a debate on the topic of renewable sources of energy versus nuclear energy. Record your findings. You may also want to consult with your Geography department as this is also an issue for them – perhaps Geography students versus D&T students!

Energy source	Process	Advantages	Disadvantages
Wind	Power of wind turns turbines, which generate electricity	• Flexibility: can be used in large-scale wind farms for national electrical grids as well as in small individual turbines for providing electricity to rural residences or grid-isolated locations • Use of wind turbines is non-polluting, environmentally friendly and sustainable; produces more than 50 times as much energy over its lifetime as is consumed by its construction and installation • Produces low-cost power if developed commercially, involving low marginal costs to run as fuel costs close to zero with relatively low maintenance costs • Could be installed offshore to minimise visual impact and take advantage of the moderate yet constant breezes	• Can only provide a small proportion of total energy needs due to the number of turbines needed in relation to space available for wind farms • Unsightly onshore wind turbines and wind farms spoil picturesque landscapes • Infrastructure required for wind farms damages landscape • Controversial – noise and vibration of moving turbine has potential to affect local community • Affects environmentally sensitive coastal sites, e.g. those with substantial birdlife
Water	Running water turns turbines and generates hydroelectric power (HEP)	• Fuel is not required so eliminates fuel costs and production of carbon dioxide • Hydroelectric plants are highly efficient with minimum running costs due to highly automated operation • Hydroelectric plants tend to have longer economic lives than fuel-fired generation, with some plants now in service having been built 50–100 years ago • High initial set-up and construction costs recovered after only a few years due to sale of vast amount electricity generated • Reservoirs created provide improved leisure and tourism, e.g. water sports, fishing, etc. • Large dams can control flooding and protect towns downstream	• Extremely expensive to construct dams and power plants • Flooding of vast areas of land to create reservoir requires local population to be relocated • Rivers may be diverted, which causes problems to local communities who rely on river for living • Dam failures (accidental or sabotage) could cause massive destruction due to flooding • Greenhouse gases produced can be high in tropical regions due to decay of plant life in reservoirs, producing methane • Hydroelectric projects can be disruptive to surrounding aquatic ecosystems, e.g. affecting fish breeding and therefore birdlife • Causes changes in the downstream river environment, e.g. erosion of river banks and low dissolved oxygen content of water
Solar	Hot water and/or electricity generated from solar energy via solar panels and photovoltaic cells	• Huge amounts of energy available from the sun • Pollution free during use • Low operating costs and very little maintenance required after initial set-up • Economically competitive especially for isolated or remote regions • Produces enough electricity for the national grid to cope with peak demand times • Local grid connected solar electricity systems can be self-sufficient	• Relatively expensive set-up costs for domestic and commercial buildings • Currently, solar electricity can be more expensive than electricity generated by other sources • Solar heat and electricity are not available at night and may be unavailable due to weather conditions, so a storage or complementary power system is required • Energy lost by converting DC current generated into AC current for use in the national grid
Biomass and biofuels	Plant materials are either incinerated to produce heat and electricity or biogas is produced from anaerobic digestion	• Relatively inexpensive source of energy • Large amounts of waste biomass materials available from agricultural processing and landfill • Production of biogas reduces the release of methane into atmosphere – a harmful greenhouse gas • By-products of biogas production can be sold and used as compost and fertiliser to improve soil condition	• Ecological damage, including deforestation and intensive farming practices • Currently, expensive processing costs of converting biomass into fuels with low yield • Incineration causes carbon dioxide pollution

Table 3.27 *Advantages and disadvantages of renewable sources of energy*

Energy source	Process	Advantages	Disadvantages
Nuclear	A controlled nuclear chain reaction creates heat, which is used to boil water, produce steam, and drive a steam turbine which in turn generates electricity	• Uses uranium, which is an abundant and widely distributed fuel • Controlled chain reaction creates heat that can also be used to heat the power station • Mitigates the greenhouse effect if used to replace fossil-fuel-derived electricity • Passively safe nuclear reactors use new technology leading to increased levels of safety to avoid leaks and overheating leading to meltdown • Future development of fission reactors, which are cleaner and more efficient	• Unpopular/mistrust with public due to media coverage of large-scale accidents, e.g. Chernobyl in 1986 • Problem of storing radioactive waste for indefinite periods, e.g. thousands of years to decay • Potential for severe radioactive contamination by accident or sabotage and proliferation of nuclear weapons in some countries • Mining of uranium causes damage to environment and pollution
Fossil fuels	The burning of hydro-carbons (oil, coal and gas) to produce heat and power	• Economies of scale: large amount of electricity produced leading to low-cost energy supply • Gas-fired power stations are very efficient • Power stations can be built almost anywhere, including dedicated transport networks assuring large quantities of fossil fuels	• Finite resources: coal, oil and gas will run out • Largest source of emissions of carbon dioxide, contributing to global warming • Generates sulphuric, carbonic and nitric acids, which cause acid rain • Fossil fuels contain radioactive materials that are also released into the atmosphere • Burning coal generates large amounts of fly ash • Mining of coal and extraction of oil and gas cause damage to the environment and pollution • Regional and global conflicts triggered over oil reserves

Table 3.28 *Advantages and disadvantages of non-renewable sources of energy*

WEBLINKS:

www.carbontrust.co.uk

The Carbon Trust helps businesses and the public sector cut carbon emissions, and supports the development of low-carbon technologies.

Responsibilities of developed countries

Global sustainable development

The challenge for people in developed countries is the need to reduce the use of scarce resources and reduce pollution. This requires a shift towards sustainable consumption and the reduction of an individual's 'carbon footprint'.

The challenges for developing countries are different. In many instances people in developing countries need to consume more, for example to gain greater access to clean water, electricity and health care. One method might be to trade more with developed countries to bring in much needed foreign investment to domestic economies. Developing countries will need to have access to markets in developed countries in order to expand. However, developed countries need to shrink their markets to address over-consumption, which creates tighter and more impenetrable markets for developing countries to sell to.

The United Nations General Assembly authorises Earth Summits where representatives of all nations meet to discuss sustainable development. Most countries have established, with the World Summit on Sustainable Development (WSSD), some form of focal point or mechanism at the national level to oversee the implementation of the Earth Summit agreements, e.g. global trade or reduction in greenhouse emissions. All countries are invited to speak at these summits. Norway, for instance, has stated several practical steps towards sustainable consumption that would include:

- improving analysis, public awareness and participation
- providing incentives for sustainable consumption
- energy: sustainable use, efficiency and renewable sources
- implementing new strategies for transportation and sustainable cities
- accelerating the use of more efficient and cleaner technologies
- strengthening international action and cooperation.

Extract from the Report of the Symposium: Sustainable Consumption, Ministry of Environment, Norway (1994)

Therefore, it follows that if global sustainable development is to succeed then all nations must firstly agree the terms and conditions and secondly, and more importantly, implement the changes needed.

THINK ABOUT THIS! 

Organise a mini Earth Summit by dividing the class into two groups representing either developed (UK or USA) or developing (Africa or China) countries. What are the main issues for global sustainable development in each type of country? How can you agree upon resolutions to tackle these problems?

Impact of industrialisation on global warming and climate change

Kyoto Protocol

The Kyoto Protocol is an amendment to the United Nations Framework Convention on Climate Change, an international treaty on the contribution of human activities to global warming. The protocol sets targets for the reduction of greenhouse gas emissions (carbon dioxide and others) by 5 per cent of 1990 levels by the nations signed up to the agreement by 2012.

The objective of the protocol is to stabilise greenhouse gas concentrations in the atmosphere at a level that would prevent dangerous changes to the world's climate. The treaty was negotiated in Kyoto, Japan in 1997 and came into force in 2005 following confirmation by Russia. As of December 2006, a total of 169 countries have signed up to the agreement, representing over 61.6 per cent of emissions from industrialised developed countries. Participating countries agreed on a set of 'common but differentiated responsibilities'.

- The largest share of historical and current global emissions of greenhouse gases has originated in developed countries.
- Per capita emissions in developing countries are still relatively low.

Figure 3.56 *Countries participating in the Kyoto Protocol*

- The share of global emissions originating in developing countries will grow to meet their social and development needs.

The United Nations Framework Convention on Climate Change (2006)

This means that developing countries are exempt from emission reduction targets in the Kyoto Protocol because they were deemed not to be the main contributors. Although developing countries such as China and India are undergoing rapid industrialisation and are exempt at present, they also share the common responsibility that all countries have in reducing emissions.

A notable exception to the Kyoto protocol is the United States which, although supporting it in principle, has never confirmed its participation. This may be partly due to the massive amounts of energy required to sustain its economy and a reluctance to accept targets imposed upon it. Instead, it has signed up to an agreement between Asia–Pacific nations including Australia, China, India, Japan and South Korea. This Asia–Pacific Partnership on Clean Development and Climate agrees to cut emissions by developing cleaner energy supplies and technologies but without the enforcement of specific targets.

Non-Fossil Fuel Obligation (NFFO)

The Non-Fossil Fuel Obligation (NFFO) was instigated in 1989 when electricity generation in the UK was privatised. Originally, money raised by the associated Fossil Fuel Levy was used to subsidise UK nuclear power generators, which continued to be state owned. However, its scope was enlarged to include the renewable energy sector in order to offer financial support for renewable technologies.

The UK Government has stated that it wishes to halve its carbon emissions by 2050, which is an ambitious task. Therefore, the government superseded the NFFO with the Renewables Obligation (April 2002) stating that all electricity suppliers should source 10 per cent of their supply from renewable technologies by 2010, rising to 15 per cent by 2015. The most promising sources of alternative energy in the UK are wind, wave and tidal.

The development of wind energy is a rapidly expanding business and the UK has the largest potential wind energy resource in Europe. It is set to account for 8 per cent of electricity generation by 2010 and is capable of producing electricity for the National Grid at prices much lower than that of coal and nuclear. The cost of wind power has reduced considerably over the last few years due to the fall in cost of turbines, the increase in size of turbines meaning that fewer are needed and decreasing project development costs as developers have gained experience. Although wind farms have been controversial when located onshore, the recent trend for offshore locations could be a compromise that will benefit the whole of the UK.

Figure 3.57 *More-favourable offshore wind farms are set to make significant contributions to the UK's renewable energy production*

Marine renewable technologies such as wave and tidal energies are a huge untapped resource for the UK, having the best wave and tidal resource in Europe. It has the potential of providing a considerable proportion of the UK's energy needs and a number of innovative marine energy devices are currently under development. However, such technologies face a number of challenges before they can become fully operational on a large-scale commercial basis.

THINK ABOUT THIS!

The development of offshore wind farms and wave and tidal technologies will affect the coastal regions of the UK. What effect could they have upon shipping and other commercial practices on which the UK relies?

WEBLINKS:

www.bwea.com

The British Wind Energy Association.

www.climatechallenge.gov.uk

Understanding climate change in the UK.

Reducing your 'carbon footprint'

A 'carbon footprint' is a measure of the impact human activities have on the environment in terms of the amount of greenhouse gases produced, measured in units of carbon dioxide. These greenhouse gases (primarily carbon dioxide, methane and nitrous oxide) are a result of industrialisation and modern living, and are contributing towards global warming.

Every time you watch television, for example, you are producing carbon emissions because of the burning of fossil fuels in the generation of electricity. Therefore, it is everyone's responsibility to reduce his or her individual carbon footprint. In the first instance this involves recognising their personal impact on global warming, including:

- annual household energy use, e.g. electricity and gas use
- annual travel, i.e. car and public transport, flights, etc.

There are many ways that an individual can save energy in the household. These include installing energy-saving lightbulbs, and turning electrical appliances off when not in use rather than using the 'standby' button. In terms of travel, car sharing, public transport, cycling or walking are viable alternatives to using the car. However, when there are very few alternatives, such as a long-haul flight, carbon offsetting might be the answer.

Carbon offsetting is way of compensating for the emissions produced with an equivalent carbon dioxide saving. These can range in scale from planting trees in the UK to conservation of wildlife habitats in Africa or South America. Some of the larger projects can also benefit local communities by providing employment, which reduces poverty.

For manufacturers, their carbon footprint can be reduced effectively by:

- applying life-cycle assessment (LCA) techniques to products in order to accurately determine the current carbon footprint
- identifying 'hot spots' in production processes in terms of energy consumption and associated carbon dioxide emissions
- optimising energy efficiency, so reducing carbon dioxide emissions and other greenhouse gases contributed from production processes
- identifying carbon offsetting solutions to neutralise the carbon dioxide emissions that cannot be eliminated by energy-saving measures.

THINK ABOUT THIS!

Use the carbon footprint calculator at www.carbonfootprint.com to work out just how much you impact global warming. What can you personally do to reduce your carbon footprint? Are there ways in which you can offset your carbon use?

WEBLINKS:

www.carbonfootprint.com

www.click4carbon.com

Both contain lots of information on reducing your carbon footprint.

Sustainable timber production

The UK relies heavily upon the import of forest products and accounts for 8 per cent of global trade in tropical hardwoods. The developing countries that produce this timber benefit little from this trade, with only 10.5 per cent of the revenue from timber production benefiting the producing country. Timber has been the focus of considerable efforts over the past decade to establish more sustainable production and trading systems. There are a number of problems associated with forests.

- **Deforestation:** the full-scale removal of forests to make way for other land uses such as settlement, infrastructure and mining. Global deforestation is currently taking place at a rate of approximately 17 million hectares each year. Deforestation occurs because trees are being felled at rates faster than new ones can be planted, and because many are cleared to make way for other land uses. When reforestation occurs it is not of equal quantity or quality as it does not replace all the benefits of the natural forest.
- **Environmental degradation of forest areas:** linked to deforestation can be soil erosion, watershed destabilisation and micro-climate change. Industrial air pollution also reduces forest health.
- **Loss of biodiversity:** deforestation and environmental degradation contribute to a rapid reduction in ecosystem, species and genetic diversity in both natural and planted forests. Forest abuse in biodiverse tropical regions is of major concern, with some scientists estimating that 1 per cent of all species are being lost each year.
- **Loss of cultural assets and knowledge:** for indigenous peoples whose lives are destroyed by deforestation. Undocumented knowledge of nurturing the forest evolving through long periods is diminishing as forest area reduces.
- **Loss of livelihood:** for forest dependent peoples, particularly in poor countries. Further social and economic problems are created elsewhere, such as in cities due to the redistribution of the local population.
- **Climate change:** both regional and global and contributes to global warming. Forests play a major role in carbon storage and with their removal more carbon dioxide enters the atmosphere.

Countries dependent upon the import of timber, such as the UK, have a responsibility to encourage the development of sustainable production and trading systems to minimise the amount of deforestation and its effects upon the environment, including:

Figure 3.58 *The Forest Stewardship Council (FSC) promotes environmentally appropriate, socially beneficial and economically viable management of the world's forests*

- no longer importing from sources that involve deforestation
- moving to supply sources in areas of ecological surplus, e.g. the high-yield plantations of Brazil, Chile and New Zealand
- certification systems that ensure that forests producing goods for the UK are sustainably managed
- timber tracing systems to ensure that products from certified forests can be identified as such
- reducing consumption through education and advisory approaches that show how to produce the same benefits from less timber
- encouraging exporting countries to make the necessary policy changes required for the transition to sustainable forest management
- supporting international efforts to control the trade in unsustainably produced wood
- improving the aid process to poor communities involved in current deforestation methods.

www.fsc.org

The Forest Stewardship Council (FSC).

ExamCafé

Teacher area

Revision summary

By now you should have a sound knowledge and understanding of a range of key topics developed over your AS level studies. You will need to build on this and develop even more in-depth knowledge and understanding of some very modern Design and Technology issues. An example is sustainability which is not simply a topic for you to revise at A2 level but a very real global issue that will affect future generations.

Again, it is important to note that the questions asked by the examiner in this exam paper will cover aspects from **all four** sections of this unit. No paper will ever focus upon one section entirely. Therefore, it is vital that you have a secure knowledge and understanding across all four sections.

You should give yourself plenty of opportunities to answer examination-style questions throughout the course so you are prepared for the final examination. Use the sample assessment materials (SAMs) and past exam papers provided by Edexcel and the questions in this handbook.

Don't forget – if in doubt, ASK! Your teacher is there to help you understand the theory in this unit. Have a good, long think about appropriate questions to ask your teacher – it may be a good idea to discuss a problem with your peers first to see if they can explain it more clearly.

Finally, keep a set of well-ordered and legible revision notes, which will help you to learn key topics and that you can always refer back to when in doubt.

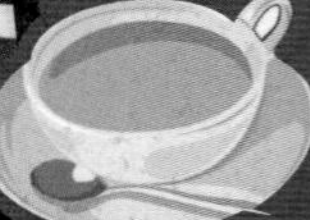

Refresh your memory

Revision checklist

- Make sure that you have answered all the questions at the end of this section.
- Make sure that your revision notes are well ordered, clear and up to date.
- Use the web links to read around each key topic so that you are well informed.
- Use SAMs and past papers to practise your exam technique.
- Discuss any problems with your peers or teacher – don't keep them to yourself!

Get the result!

Tips for answering questions

Questions at A2 level will be much more difficult than those at AS level so it is extremely important that you read each question carefully before you respond. It might be a good idea to use a piece of scrap paper to outline your response if you think you have enough time.

Always look at the amount of marks awarded for each question in brackets. This will give you a good indication of how many points need to be raised in your response. As a general rule of thumb, look at the following command words and what you have to do in order to gain the marks:

Give, state, name	(1 mark)	These types of questions will not feature heavily at A2 level but may appear at the beginning of the paper, or question part. They are designed to ease you into the question with a simple statement or short phrase.
Describe, outline	(2+ marks)	These types of questions are quite straightforward. They ask you to simply describe something in detail. Some questions may also ask you to use notes and sketches; you can gain marks with the use of a clearly labelled sketch.
Explain, justify	(2+ marks)	These types of questions will be commonplace in this exam. They are asking you to respond in detail to the question – no short phrases will be acceptable here. Instead, you will have to make a valid point and justify it.
Assess, consider, discuss	(3+ marks)	These types of questions require a more detailed response. You need to structure your own response, addressing a number of key points (possibly advantages and disadvantages) in your answer, sometimes coming up with a resulting argument or conclusion.
Compare	(4+ marks)	These types of questions will normally give you two different products and ask you to compare them either in terms of their design and manufacture or issues relating to sustainability.
Evaluate	(4+ marks)	These types of questions will appear towards the end of the paper or question part and are designed to stretch and challenge the more able student. They require you to make a well-balanced argument, usually involving both advantages and disadvantages.

Ask the examiner: Worked examples

The following four questions should demonstrate the style of questions using some of the different types of command words. The places where marks have been awarded are indicated in brackets. These are referred to as 'trigger points' and are parts of the examiner's mark scheme where marks are expected to be awarded.

Exam question 1

Explain **two** advantages to the **manufacturer** of using an electronic point of sale (EPOS) system to gather sales information.

(4 marks)

Martin

1. *It is faster at collecting sales information.*

(1 mark)

2. *It uses a barcode to identify the product.*

(1 mark)

Two statements are offered, neither of which are justified. The first statement is relevant, as faster collection of sales information would benefit the manufacturer, but why? The second statement simply notifies us of a component of EPOS and not an advantage.

David

1. *EPOS provides an easier and faster way of collecting sales information* **(1 mark)** *which can be used by the manufacturer to respond quickly to consumer demand* **(1 mark)**.
2. *EPOS enables the manufacturer to operate a JIT system* **(1mark)** *which reduces the need for large stock levels and so reduces costs* **(1mark)**.

(4 marks)

Two fully justified responses are given that clearly demonstrate the candidate's knowledge and understanding of EPOS when applied to the **manufacturer**. The focus of this type of question could quite easily be on the retailer or the consumer so it is really important that you read the question carefully.

Exam question 2

Discuss the benefits of using genetic engineering in the production of timber

(3 marks)

Joseph

Genetic engineering benefits the production of timber *because it produces quicker-growing trees that are more resistant to disease* **(1 mark)** *so there is a plentiful supply of raw materials for things like doorframes* **(1 mark)**.

(2 marks)

The first part of the response contains two relevant points: 'quicker-growing trees' and 'more resistant to disease', but they are not justified. The second part of the response seems to pull the whole response together by making reference to there being trees grown specifically to make doorframes.

Leena

Genetic engineering can be used to produce trees with enhanced aesthetics **(1 mark)**. *This means that it may not need as much painting or polishing* **(1 mark)**. *The growing of genetically modified trees also involves better forest management, which reduces the problems of deforestation* **(1 mark)** *caused by the vast amount of raw materials needed for the timber industry.*

(3 marks)

This is a good response that looks at the question from a sustainable point of view. The candidate makes the connection between the need for vast amounts of raw materials and how genetic modification of trees can be used in a positive way to supply the needs of the timber industry.

Exam question 3

Aesthetic design movements throughout history have influenced the styling of products and architecture. Outline the 'style' of **one** of the following aesthetic design movements:

- Art Nouveau
- Art Deco

(6 marks)

Lizzie

Art Nouveau was a movement that used flowing lines in its designs as opposed to Art Deco, which used zig-zag shapes. It was a time when people had more money and could afford to buy more products. They wanted things that looked good so they liked the idea of more stylised decoration such as flowing lines that looked like plants **(1 mark)**. *Women featured heavily in their designs with long, flowing hair* **(1 mark)**. *Designers were also inspired by Japanese culture.*

(2 marks)

You will see that the first sentence is not awarded a mark as the question asks students to outline the style of **one** movement only and not to compare two. The second sentence is simply waffle. The following two sentences gain marks because they describe two styling points whilst the last sentence simply makes a statement with no explanation.

Jon

The style of the Art Nouveau movement is probably best characterised by the languid line **(1 mark)**. *Designers found inspiration in natural forms* **(1 mark)** *and represented them with curvy 'whiplash' lines and stylised flowers* **(1 mark)**. *The peacock feather motif was often featured because it symbolised the hedonistic views of the time* **(1 mark)**. *Art Nouveau was often referred to as 'feminine art' due to its frequent use of languid female figures with long, flowing hair* **(1 mark)**. *The grid structures of Japanese interiors provided vertical lines and height to many pieces of furniture,* **(1 mark)** *especially those of Charles Rennie Mackintosh.*

(6 marks)

When you read this response you can instantly tell that this student is well informed on the subject of Art Nouveau and has studied it in some detail. Again, the marks in brackets indicate where the marks have been awarded. Here, there are clearly six different aspects to Art Nouveau styling succinctly presented.

Exam question 4

Evaluate the use of fully automated production and assembly lines incorporating robots when manufacturing products compared with labour-intensive methods. **(10 marks)**

Mark

Machines can do things quicker, better and more efficiently than humans. A robot can work for 24 hours a day without a break or getting tired but a human can't **(1 mark)**. *It also costs a lot to employ many workers when you don't have to pay a machine anything except that it costs a lot to buy in the first place* **(1 mark)**. *People that work on production lines often become bored with their jobs as they do the same thing every day and they are usually not very well paid* **(1 mark)**.

(3 marks)

The candidate has started the response with a very basic statement that does not gain any marks at A2 level. The response focuses upon the idea that machines are more efficient than humans but fails to justify many of the points raised sufficiently. The candidate would have benefited perhaps from structuring the response in rough first in order to fully evaluate the topic.

Justin

Advantages of fully automated production over labour-intensive methods:

- *Automation increases productivity and reduces running costs* **(1 mark)** *due to the efficiency of machines and less wages paid to a large manual workforce* **(1 mark)**.
- *Robots can work freely in hazardous conditions such as paint shops,* **(1 mark)** *which humans could not do without risks to health and safety* **(1 mark)**.
- *Automation can free up the workforce from repetitive manual labour* **(1 mark)** *allowing more people to enter higher-skilled jobs, which are typically higher paying* **(1 mark)**.

Disadvantages of fully automated production over labour-intensive methods:

- *Automated processes with robots are not suitable for extremely detailed manufacturing and finishing activities* **(1 mark)** *where the human senses such as vision, touch and pattern recognition are still better* **(1 mark)**.
- *Robots do not have the ability to learn and make decisions when the required data does not exist* **(1 mark)** *whereas a human workforce can react quickly to change and adapt accordingly* **(1 mark)**.

(10 marks)

Here, the candidate has structured the response in a logical manner as 'evaluate' questions ask for both advantages and disadvantages. The bullet-pointed responses under each heading are fully justified and demonstrate a very good understanding of the impact of automation upon manufacturing and employment. Note that more responses are given for advantages than disadvantages – this is perfectly ok as long as both are present.

Practice questions

1. Explain **two** advantages of using virtual modelling and testing in the development of a new product. **(4 marks)**
2. Explain **three** ways in which computer-integrated manufacturing (CIM) benefits a manufacturer mass producing a product. **(6 marks)**
3. Discuss the impact of the development of industrial mass-production on:

   (i) workers. **(3 marks)**

   (ii) consumers. **(3 marks)**
4. Discuss the effects of using biodegradable polymers, such as Biopol®, on the packaging industry. **(6 marks)**
5. Outline the use of **one** smart material for an innovative application. **(4 marks)**
6. Discuss the relationship between ergonomics and anthropometrics. **(4 marks)**
7. Consider the ergonomic factors involved when designing a bicycle for an adult user. **(6 marks)**
8. Figures 1 and 2 below show two different lemon squeezers.

   The lemon squeezer in Figure 1 is called 'Juicy Salif', designed by Philippe Starck, and Figure 2 shows a widely available lemon squeezer from a high-street shop.

   Figure 1

   Figure 2

   Compare the lemon squeezer in Figure 1 with the lemon squeezer in Figure 2 with reference to the following:

   (i) form.

   (ii) function. **(8 marks)**
9. Evaluate the use of alternatives to fossil fuels in the transportation of products from the manufacturer to the retailer. **(8 marks)**
10. Evaluate the impact of recycling upon sustainable product design. **(8 marks)**

Unit 4:

Commercial Design

Summary of expectations

1 What to expect

In this unit, you are given the opportunity to apply the skills that you have acquired and developed throughout this course of study to design and make a product of your choice that complies with the requirements of a Resistant Materials project. You are encouraged to be creative and adventurous in your work. Throughout this course you are expected to take ownership of all aspects of your work, and to take total control of your responses with your teacher as facilitator. You are strongly advised to target assessment criteria effectively in order to maximise your achievements.

In order to reach high attainment levels, you must adopt a commercial design approach to your work, reflecting how a professional designer might deal with a design problem and its resolution. The choice of design problem should have a real commercial use in that it should be useful to a wider range of users and not simply be designed for yourself. The product should be designed to be commercially manufactured. This can be batch or mass produced, unless it has been specifically commissioned as a 'one-off'.

The design problem should provide opportunities for a client, or user group, to have input into decision making at various stages of the design and make process. A client/user group is defined as any third party you identify that is referred to and that can give informed critical feedback at various stages throughout the design process. Clients/user groups do not need to be specialists or experts. They can be drawn from any relevant group of people and may include other students, friends or family members.

A key feature of this unit is for you to consider issues related to sustainability and the impact your product may have on the environment. You may choose to design and make a sustainable product but, if you do not, you should still consider the issues of sustainability at relevant points in your designing and making activities. Sustainable issues could include materials production and selection, manufacturing processes, use of the product as well as its disposal or recycling.

This unit is set and marked by your teachers, then sent to Edexcel for moderation (sampling and checking of teachers' marks).

2 What is a Resistant Materials project?

A Resistant Materials project requires you to work with a range of materials and manufacturing processes in order to meet specific design issues and overcome difficulties. At this level your project should be focused on the needs of a client/user group and they will have a direct input into the decision making at various stages of your project. The solution should be one which can be fully tested, and not a model.

3 How will it be assessed?

The coursework requirement at A2 Level is a full design and make activity, offering you the opportunity to demonstrate the knowledge, skills and competencies that you have gained from your AS studies. A breakdown of each assessment criterion statement will be outlined in the *To be successful you will* sections of this student book.

Where large numbers of marks are assigned to assessment sections, such as design and development and making, these have been broken down into smaller sub-sections to allow clearer and easier access to marks.

The maximum number of marks available for this unit is 90.

Section	Sub-section		Marks
Product design and make	A. Research and analysis		4
	B. Product specification		6
	C. Design and development:	Design	10
		Review	4
		Develop	10
		Communicate	6
	D. Planning		6
	E. Making:	Use of tools and equipment	9
		Quality	16
		Complexity/ level of demand	9
	F. Testing and evaluating		10
	Total marks:		90

Please note: It is extremely important that you sign the authentication statement in your Candidate Assessment Booklet (CAB) before your work is marked. If you do not authenticate your work Edexcel will give you zero credit for this unit.

4 The coursework project folder

This unit results in the development of an appropriate product supported by a design folder. The folder, which should include information and communication technology (ICT) generated images where appropriate, can only be submitted on A3 paper and is likely to be no more than 30 pages long. You can also submit your work electronically for moderation provided it is saved in a format that can be easily opened and read on any computer system, i.e. a PDF document.

Your product design and make folder must be organised in a clear and logical manner that reflects the order of the assessment sections. It is important that each page is evidenced in the appropriate section. This will allow your teacher to easily mark your work and provide your Edexcel moderator with a clear indication of your skills and ability.

Suggested contents		Suggested page breakdown
Title page with specification name and number, candidate name and number, centre name and number, title of project and year of submission		Extra page
Contents page		Extra page
A. Research and analysis		3–4
B. Product specification		1–2
C. Design and development:	Design	3–4
	Review	1–2
	Develop	3–4
	Communicate	(Evidenced throughout section) Working drawings 1–2 Pictorial drawings 1–2
D. Planning		2–3
E. Making:	Use of tools and equipment	3–4
	Quality	
	Complexity/level of demand	
F. Testing and evaluating		2–3
Bibliography		Extra page
Total pages:		20–30

5 How much is it worth?

The product design and make coursework project is worth 60 per cent of the A2 course and 30 per cent of the overall full Advanced GCE.

Unit 4	Weighting
A2 level	60%
Full GCE	30%

Product design and make

Getting started

In order to get this unit started, you must first identify a specific need, or problem, and derive from it a detailed design brief. Don't forget that in this unit you must adopt a commercial design approach, acting like a professional designer. Therefore, your choice of design problem should have a real commercial use and involve a real client or specific user group.

Problem/need	Client/user group	Outcome(s)
Your cousin has now grown out of his baby cot and his parents are looking for a new novelty themed junior bed.	**Client:** aunt and uncle **User group:** cousin	Custom made bed on a novelty theme
Your local feeder primary school needs to improve the reading of their Year 3 and Year 4 students	**Client:** headteacher/ teacher in charge of library **User group:** primary school students	Book case/ display stands
Your dad's company is looking to expand their range of garden furniture	**Client:** your dad and other members of his company	Garden seating/ tables/swing seats/hammocks

Table 4.1 Ideas for commercial design projects

Design brief

To design and make a new book case display system to encourage the Year 3 and Year 4 students to read more and to encourage more borrowing of school library books.

I intend to design a storage system to display the books more readily and to make them more appealing to the Year 3 and Year 4 students. At the moment they are simply stacked on a shelf with all the covers turned against each other and the bright attractive covers are all hidden from sight.

Clients: Mr Hill, headteacher; Ms Jenkins, teacher in charge of library

User groups: Year 3 and Year 4 students and classes

Figure 4.1 *An example of an appropriate design brief*

ACTIVITY:

Take a long, hard look around a primary school site. Identify **ten** aspects of the site, building or classroom facilities that could be improved. This could include where students can store their wet coats if they have been outside in the rain, how teachers can display the library books to encourage more reading, what the outside playground facilities are like or whether there is disabled access into the buildings.

Produce a report entitled *Ten things that could be improved.* Compile the top ten in order of the schoolchildren's priorities and compare these with your classmates, with the number one spot being something that potentially could be a good project.

There are many opportunities for gaining client feedback throughout this project from site staff, teachers and even the headteacher, as well as user group feedback from the students themselves.

Ignore headings + marks.

A. Research and analysis (4 marks)

Once you have identified an appropriate need and written a detailed design brief, you must analyse the need in order to focus on the research needed to help your work progress. You should use a range of research strategies to gather useful and relevant information that will help with your designing and making activities.

When gathering information, it is important that you are clear about what you need to find out. Research should be highly selective, ensuring that the information gathered is useful and relevant to the client/user group's needs which were identified and finalised during analysis. Research should be focused and succinct and contain no worthless padding. Avoid downloading large amounts of information from the Internet, or cutting and pasting from catalogues and databases, without providing detailed annotation to explain the selected information. All sources must be fully acknowledged to avoid the risk of plagiarism.

A good starting point for research and analysis could include an interview, or discussion, with the client/user group to establish their thoughts and preferences regarding the proposed product. This information should be used to guide your analysis and research activities. In the analysis, you should ensure that you focus closely on the identified need, avoiding any general statements that are of no use and could be applied to any design situation.

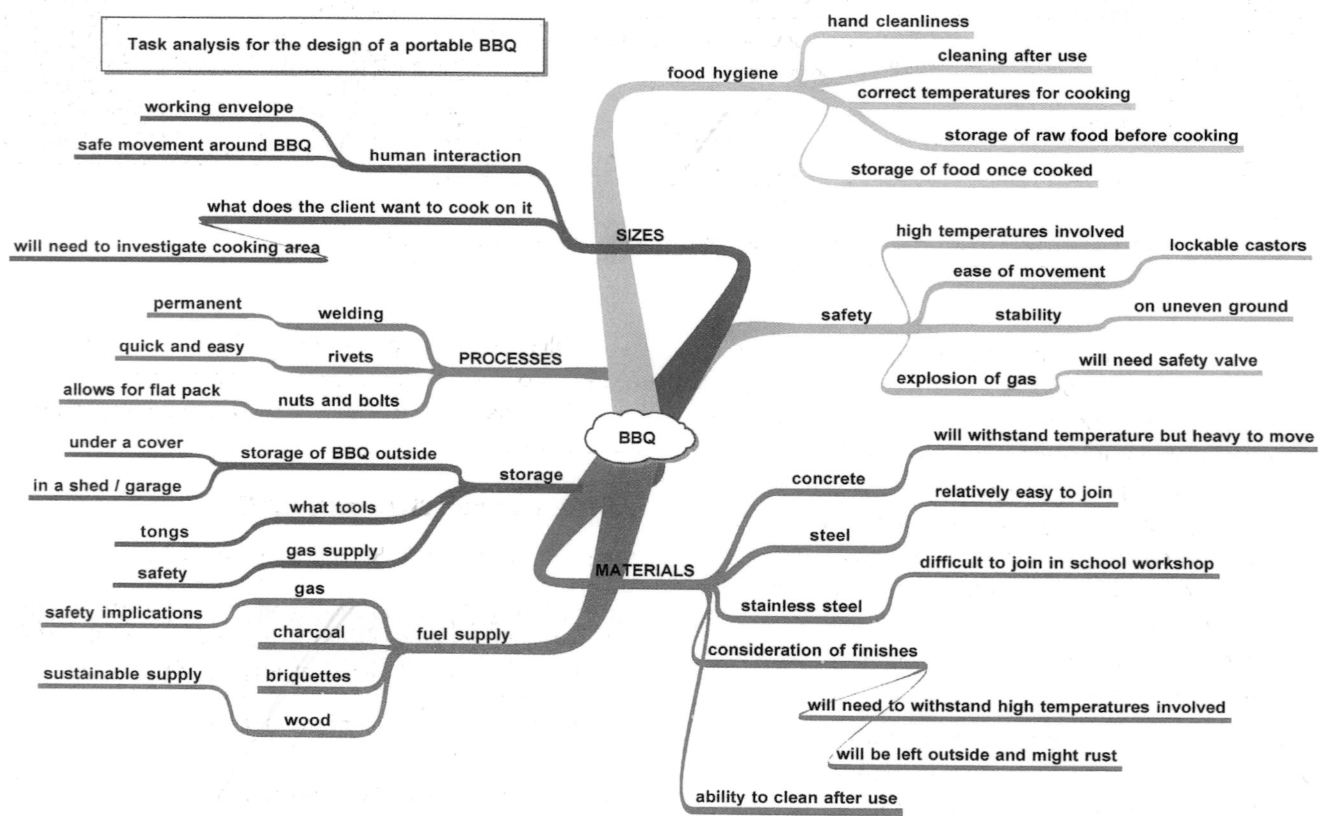

Figure 4.2 *This student has explored the task in detail by analysing the key words in the form of a spider diagram*

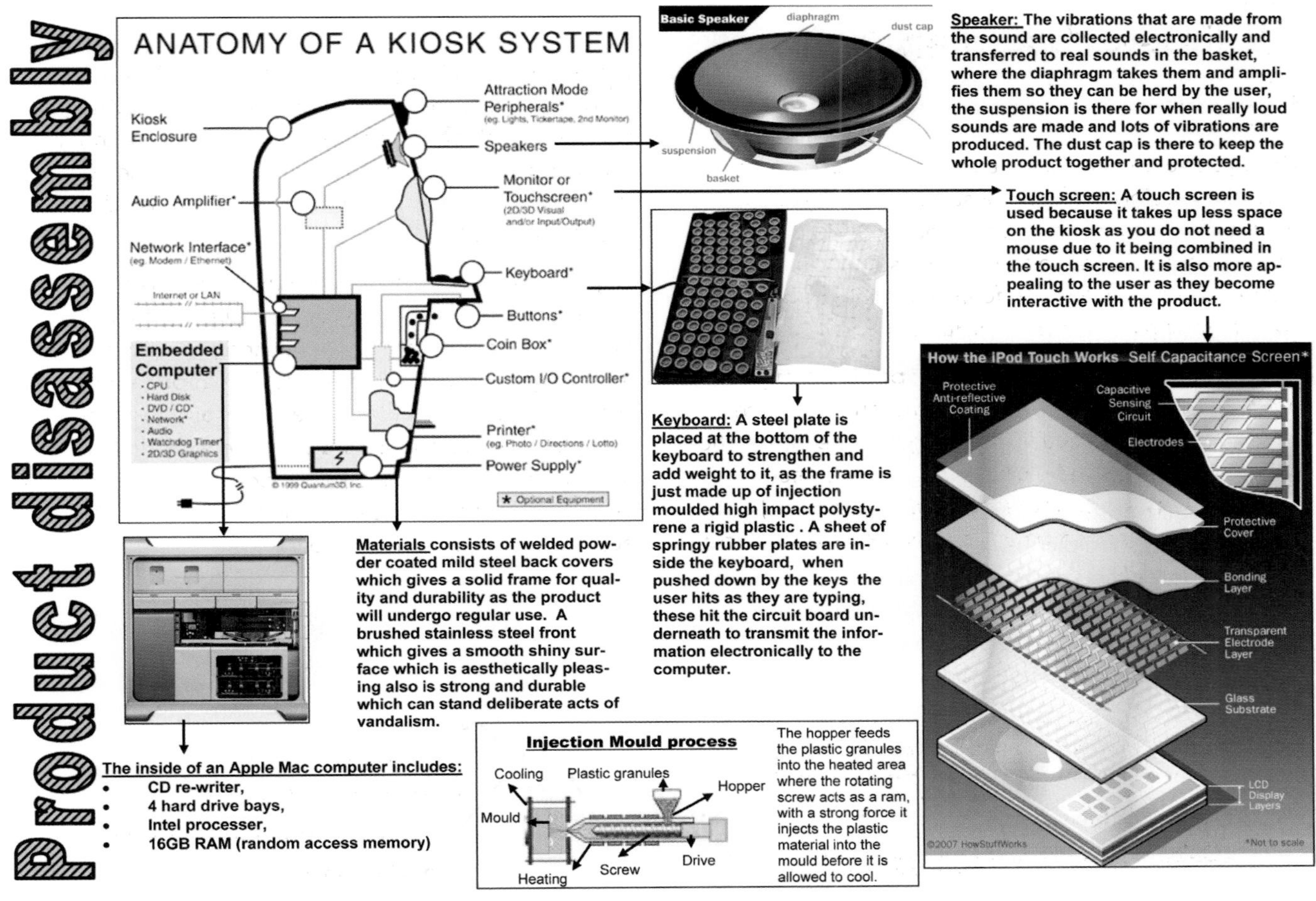

Product disassembly
ANATOMY OF A KIOSK SYSTEM
Kiosk Enclosure
Audio Amplifier*
Network Interface*
Internet or LAN
Embedded Computer
Attraction Mode Peripherals*
Speakers
Monitor or Touchscreen*
Keyboard*
Buttons*
Coin Box*
Custom I/O Controller*
Printer*
Power Supply*
Optional Equipment
Basic Speaker
diaphragm
dust cap
suspension
basket
Speaker: The vibrations that are made from the sound are collected electronically and transferred to real sounds in the basket, where the diaphragm takes them and amplifies them so they can be herd by the user, the suspension is there for when really loud sounds are made and lots of vibrations are produced. The dust cap is there to keep the whole product together and protected.
Touch screen: A touch screen is used because it takes up less space on the kiosk as you do not need a mouse due to it being combined in the touch screen. It is also more appealing to the user as they become interactive with the product.
How the iPod Touch Works Self Capacitance Screen*
Protective Anti-reflective Coating
Capacitive Sensing Circuit
Electrodes
Protective Cover
Bonding Layer
Transparent Electrode Layer
Glass Substrate
LCD Display Layers
Keyboard: A steel plate is placed at the bottom of the keyboard to strengthen and add weight to it, as the frame is just made up of injection moulded high impact polystyrene a rigid plastic . A sheet of springy rubber plates are inside the keyboard, when pushed down by the keys the user hits as they are typing, these hit the circuit board underneath to transmit the information electronically to the computer.
Materials consists of welded powder coated mild steel back covers which gives a solid frame for quality and durability as the product will undergo regular use. A brushed stainless steel front which gives a smooth shiny surface which is aesthetically pleasing also is strong and durable which can stand deliberate acts of vandalism.
The inside of an Apple Mac computer includes:
CD re-writer,
4 hard drive bays,
Intel processer,
16GB RAM (random access memory)
Injection Mould process
Cooling
Plastic granules
Hopper
Mould
Screw
Drive
Heating
The hopper feeds the plastic granules into the heated area where the rotating screw acts as a ram, with a strong force it injects the plastic material into the mould before it is allowed to cool.

Figure 4.3 Using skills developed from Unit 1 to disassemble and analyse a relevant existing product

Research could include the analysis of existing similar products to find out about materials, processes and construction methods used in commercial manufacture. Market research will allow you to test the viability of your intended product beyond the needs of the client/user group. Surveys, or questionnaires, should be designed carefully, avoiding questions that are general, and useless, in helping with the design process. A questionnaire should not be included simply for the sake of doing so; its use and the questions asked within it should be justified.

When researching materials, components and processes, you should take into consideration the concept of sustainability so that you are able to make responsible and informed decisions about the impact of materials and resources upon the environment.

When all information gathering has been completed, you should analyse your research in order to help write a product specification that is relevant, meaningful and measurable.

To be successful you will:

Assessment criteria A. Research and analysis

Level of response	Mark range
Produce detailed analysis with most design needs clarified. **(1 mark)** Present selective research that focuses on the needs identified in the analysis. **(1 mark)**	3–4
Produce limited analysis with some design needs clarified. **(1 mark)** Present superficial research that does not focus on the needs identified in the analysis. **(1 mark)**	1–2

B. Product specification (6 marks)

It is important that you develop, and write, a detailed specification as it will be used throughout the design process to review your ideas and their development. It will also be used to check that the design requirements and client/user group needs are being satisfied. The specification should be used as a basis for testing and evaluating the completed product and any future modifications suggested should be referenced to specification criteria in order to check the success of your final product.

The starting point for a successful specification should be after the research when essential requirements have been established as a result of studying the information gathered. You should consult with your client/user group to agree the specification points and to ensure that the criteria meet their needs. When specifying materials, components and processes you should consider sustainability and make decisions based on the environmental costs of extracting and processing the selected materials and the product manufacture, lifespan and disposal.

When writing a specification, you should try to avoid a rambling collection of points. The specification should be informed by your research findings. An effective specification is organised logically and could be achieved by using sub-headings such as:

- **purpose** – what is the aim or end-use of the product?
- **form** – what shape/style must the product take?
- **function** – what must the product specifically do?
- **user requirements** – what qualities must the product have to make it attractive to client/user group?
- **performance requirements** – what technical considerations need to be achieved within the product?
- **materials and components** – what materials and components should be used to aid performance?
- **size** – what physical dimensions are required?
- **safety** – what factors need to be considered to make the product safe to use?
- **quality** – how can a high-quality product be assured?
- **scale of production** – how many are to be made and by what manufacturing processes?
- **cost** – what are the considerations in determining cost?

Each specification point should contain more than a single piece of information, so that each statement is fully justified by giving a reason for the initial point. It is not sufficient to say 'the material used is medium density fibreboard (MDF)', as this is not justified until 'because it is available in large flat sheets and is ideal for veneering' is added.

Specification points should be technical and measurable where possible, so that testing and evaluation can be realistic. It is extremely important that your specification points are not superficial or general.

ACTIVITY:

Using the example of the garden bench design specification in Figure 4.4, outline the tests that could be made to determine whether each specification has been matched. This could include sufficient length to accommodate four seated people and anthropometric height for the average user.

This is measurable as you could easily test how much space a seated person needs to occupy to be safe and comfortable. You can then obtain the minimum length of the garden bench by multiplying this figure by four. Anthropometric data charts can be referred to in order to determine a height suitable for the majority of the population (95th percentile).

To be successful you will:

Assessment criteria B. Product specification

Level of response	Mark range
Compile specification points that are realistic, technical and measurable. **(1 mark)** Produce a specification that fully justifies points developed from research in consultation with a client/user group. **(1 mark)** Realistically consider the sustainability of relevant resources when developing specification points. **(1 mark)**	4–6
Compile specification points that are realistic but not measurable. **(1 mark)** Produce some specification points that are developed from research in limited consultation with a client/user group, but are not justified. **(1 mark)** Superficially consider the sustainability of resources when developing specification points. **(1 mark)**	1–3

Design brief
To design and make a garden bench for the headteacher's lawn

Client: Mr. Chamberlin, headteacher
User group: Headteacher and parents of prospective pupils

Product specification
The garden bench must:

Purpose
- Provide a safe, comfortable area on which to sit whilst looking out over the neat, well-kept school gardens.

Form
- Have a traditional style to reflect that of the Victorian buildings surrounding the gardens.

Function
- Provide a comfortable seating space for four people to sit on, since the headteacher takes the parents of prospective pupils into the garden area.
- Provide a shaded hood to protect the people sitting on the bench from the sun so that they do not get burnt or too hot.

User requirements
- Provide a stable seat which could be placed on a variety of surfaces, since there are a number of different surfaces in the garden such as grass, tarmac and a paved area.
- Provide sufficient shelter from direct sunlight so that people sitting on it do not get too hot.

Performance requirements
- Not wobble when placed on various surfaces so that it does not become unsafe to use.
- Provide a back support to lean against so it is comfortable, because users may sit on it for some time.
- Support the weight of four adults (maximum of 400kg evenly distributed) so it is safe to sit on without risk of it collapsing.
- Be a maximum of 420mm high, which will comfortably seat the 95th percentile.

Materials and components
- Be manufactured using traditional English hardwood so that it is in keeping with the window and door frames that lead onto the garden.
- Be finished to protect the material from sunlight deterioration and from rain and wet or damp conditions.

Size
- Be easy to lift and move so that no injury will be caused to those moving it.
- Be a minimum of 1.7 metres long so that it can accommodate four seated people.

Safety
- Not have any sharp corners or edges that could cause injury.
- Not weigh more than 50kg, which will make it a relatively safe weight for two people to move.

Quality
- Use materials and components (adhesives / fittings) that have been subject to British Standard testing, so that they are known to be reliable and have been subjected to safety standard tests.
- Be manufactured using appropriate quality control (QC) procedures in order to assure a high-quality outcome.

Scale of production
- Be able to be manufactured as a one-off, since only one is required by the headteacher.
- Be able to be manufactured using efficient fabrication techniques with use of standard components where necessary.

Cost
- Come in on budget at £500 whilst still meeting the requirements of the client.

Figure 4.4 *A sufficiently detailed product specification that would enable a student to effectively review design ideas and test the final outcome*

C. Design and development

FACTFILE:

- This *Design and development* section carries large numbers of marks but is sub-divided into four areas – design, review, develop and communicate – which are explained individually to show what should be presented as evidence to gain marks.
- An important feature of this section is that you should consider issues related to sustainability and the impact your product might have on the environment.

Design (10 marks)

In this section, you have the opportunity to apply your design skills and the advanced knowledge of materials, components, processes and techniques developed through your experience of the AS units. Design ideas should be produced that are realistic, workable, and that address the needs identified in your specification. Designs should be annotated and include as much detail as possible of materials, components, processes and techniques that could be used to construct each design idea.

A suitable starting point for design ideas might be to firstly explore shape and form. These aesthetic considerations are the most obvious visual reference by which you can begin to add more detail to your design ideas. If you decide upon a suitable shape then you can begin to consider how it could be made and what materials and components are necessary. If you simply explore the aesthetics of your product, or environment, then you will not be able to access the higher marks. Remember that your ideas must be workable, so a sufficient degree of detail is necessary to communicate your intentions and test whether or not they can be developed further.

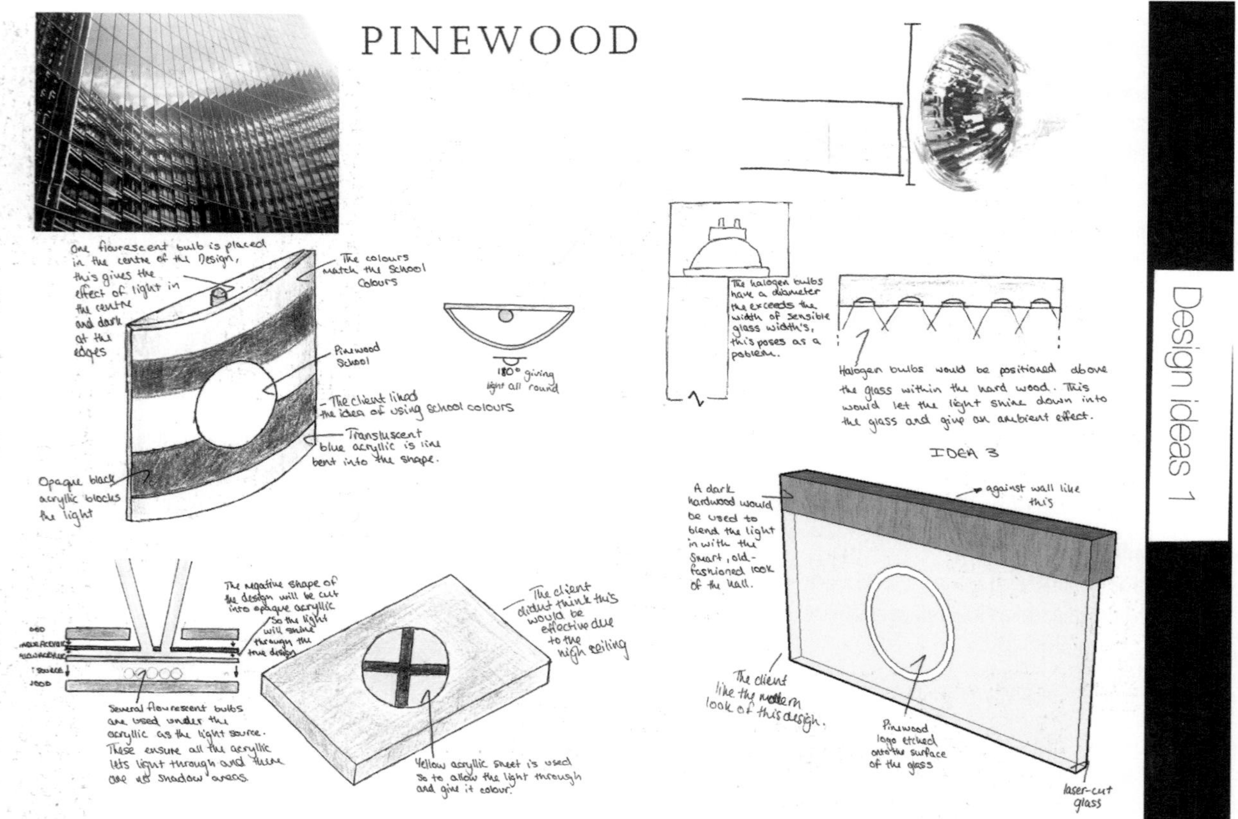

Figure 4.5 A student's visual design sheets explore ideas in detail using external stimuli for inspiration

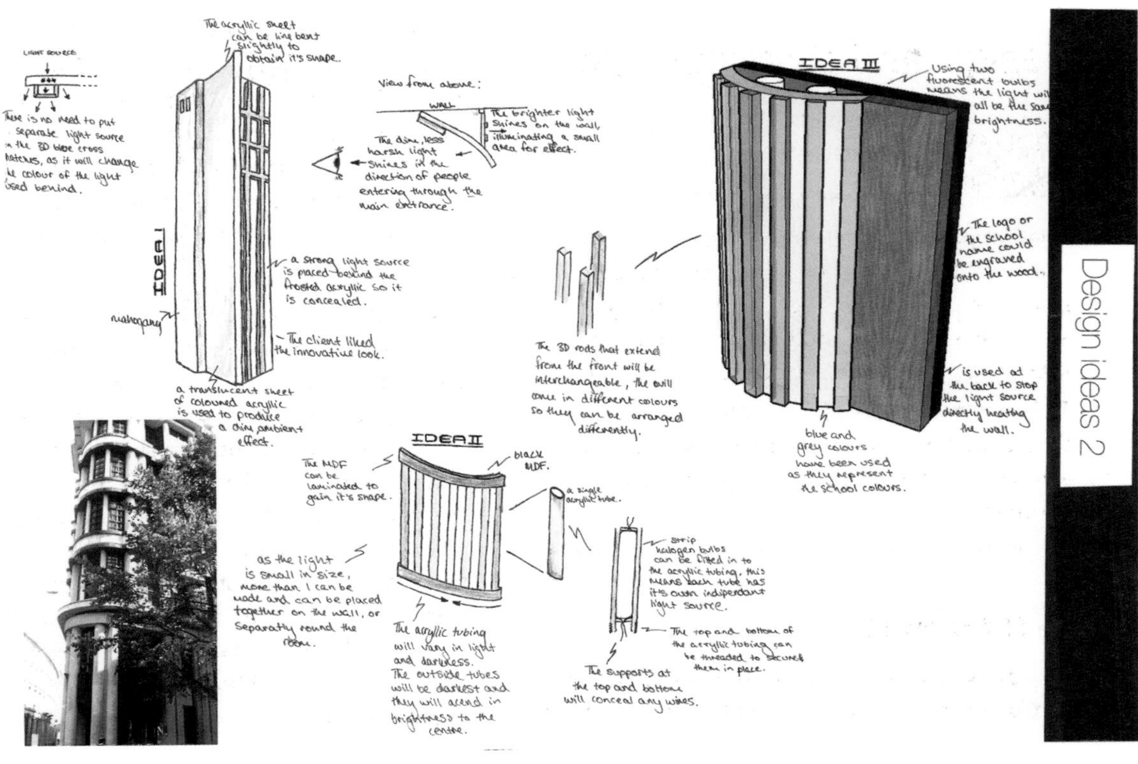

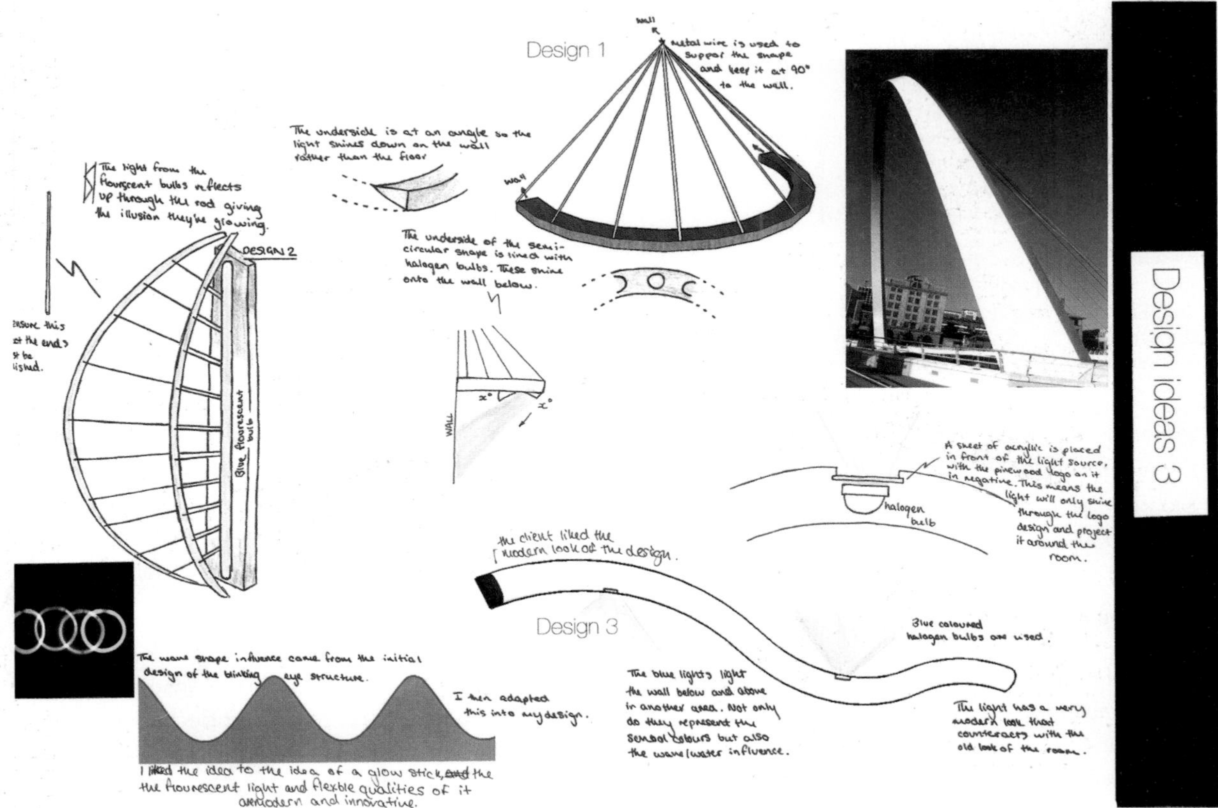

Figure 4.5 A student's highly visual design sheets explore ideas in detail using external stimuli for inspiration (continued)

ACTIVITY:

Once you and your classmates have completed a series of design sheets for your project:

- Swap folders with the person next to you and study their designs in detail in order to carry out a critique of their work.
- Make a list of things that are not communicated as well as they could be, for example a sketch may be unclear and requires additional annotation to explain a concept.
- Make a list of problems that you can see with some of their designs, for example the material chosen may not have sufficient properties for its application. Suggest alternatives.
- Politely feed back your observations using constructive criticism and identifying areas for further improvement.

Once this activity is complete and everyone has had some feedback, the necessary amendments should be made to your project. An activity like this can have a positive effect upon your project as feedback is coming from your friends rather than from your teacher.

To be successful you will:

Assessment criteria C. Design and development: Design

Level of response	Mark range
Present alternative ideas that are realistic, workable and detailed. **(1 mark)** Demonstrate a detailed understanding of materials, processes and techniques in ideas that are supported by research information. **(1 mark)** Address all specification points in ideas. **(1 mark)** Show client/user group feedback. **(1 mark)**	7–10
Present alternative design ideas that are realistic and workable. **(1 mark)** Use relevant research to produce detailed ideas. **(1 mark)** Address most specification points in ideas. **(1 mark)**	4–6
Present alternative design ideas that are similar and simplistic. **(1 mark)** Use limited research to produce similar ideas. **(1 mark)** Address a limited number of specification points. **(1 mark)**	1–3

Review (4 marks)

An important part of your designing is to review and objectively evaluate your design ideas as they are produced. The comments made in reviewing design ideas should be based on objective, formative evaluation of each idea and should always be referenced to the specification in order to check the idea's potential in fulfilling your client/user group's need. It is always good practice to choose more than one idea for a detailed review as this gives the client/user group a choice of suitable outcomes.

Please note that you should not use simple 'tick-boxes' when reviewing design ideas, as this is always subjective and worthless in evaluating ideas against a specification effectively. In addition, avoid simple Yes/No answers as they do not allow any useful decisions to be made when deciding whether specification points have been met. Remember that all specification points need to be justified.

Do not feel that all of your specification points have to be successfully satisfied at this stage. No design is ever perfect first time around and that is why large amounts of money are used in the development of any product. Initial design ideas should address the majority of the criteria but it is at the development stage that you will have to ensure that all specification points are met. At this review stage you are simply pointing out areas for further development.

As part of your review, design ideas should be discussed with your client/user group to ensure, through feedback, their suitability for their intended purpose. Information should be communicated through logical and well-organised statements, using specialist technical vocabulary. In addition you should consider, and justify, some of their design decisions with reference to sustainability.

ACTIVITY:

Choose your best two design ideas, photocopy them and present them on a board for your client/user group to review. Prepare a short presentation of each idea, outlining the concept behind the idea, and be prepared to answer any questions that your client/user group may have.

Ask your client for their opinions on both designs – what they like/dislike and what they would like to see in any future designs. It might be a good idea to have prepared a brief questionnaire to prompt the discussion and record their views. Alternatively, conduct an interview using a Dictaphone to record the conversation so that you don't miss any relevant information.

To be successful you will:

Assessment criteria C. Design and development: Review

Level of response	Mark range
Present objective evaluative comments against most specification points that consider client/user group feedback. **(1 mark)** Include evaluative comments on realistic issues of sustainability relating to design and resources. **(1 mark)**	3–4
Present general and subjective comments against some specification points. **(1 mark)** Superficially evaluate an aspect of sustainability. **(1 mark)**	1–2

Develop (10 marks)

In this section, you will develop a final design proposal in consultation with your client/user group. Development of the final design proposal will give you the opportunity to bring together the best and most appropriate features of your initial design ideas. This refined final design proposal should not only meet all of the requirements of your product specification but also satisfy your client/user group needs.

You must show the development of your design, demonstrating how it has changed and moved on from initial ideas, using the results of review/ evaluation and client feedback. It is not good practice to simply take an initial idea, make superficial or cosmetic changes, and then present it as a final developed proposal.

You should include as much detailed information on all aspects of the developed design as possible, including technical details of materials and components and their selection, processes and techniques. This is an opportunity for you to demonstrate an advanced knowledge and understanding of design and make activities.

Modelling should be used to test features such as proportions, scale, function, sub-systems, etc. Modelling can be achieved through the use of traditional materials, or 2D and/or 3D computer simulations. Evidence of modelling should be presented through clear, well-annotated photographs. Consultation with the client/user group should be evidenced in order to justify and clarify final design details.

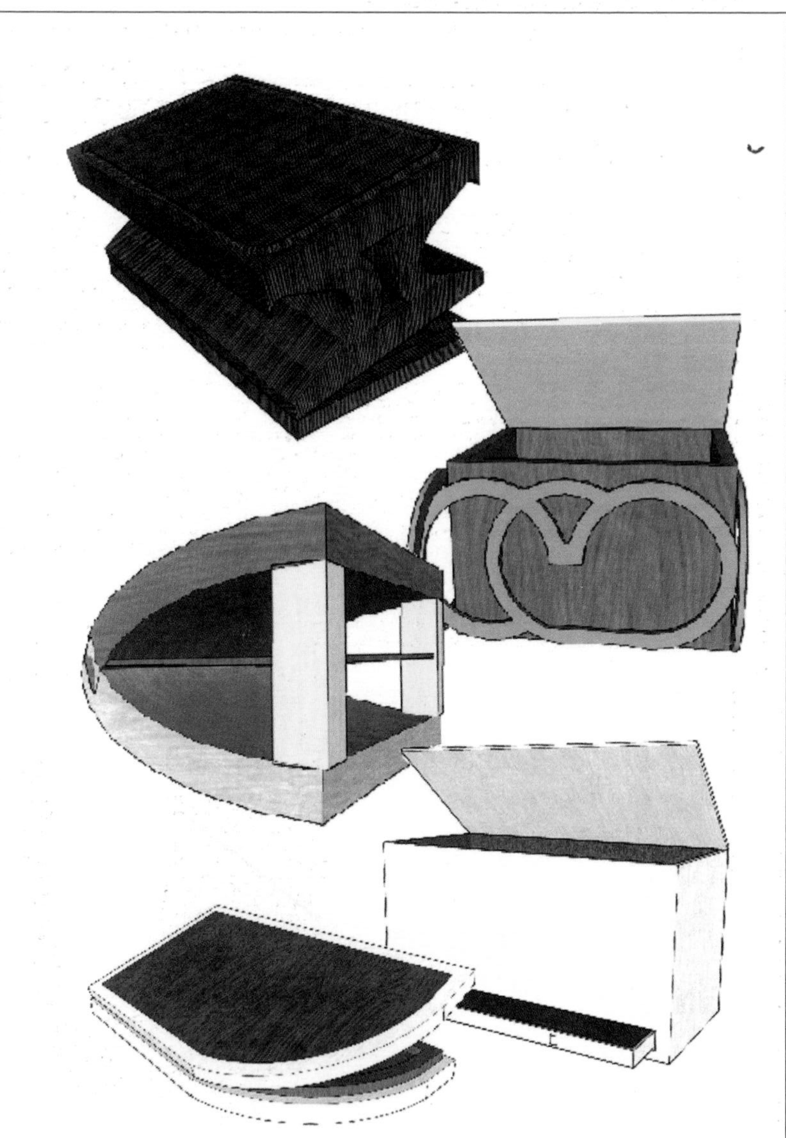

From the detailed Evaluation of Designs I have seen a varied selection of Ideas. Some of the concepts conform to the specification points more than others. From the individual analysis of each design, I can evaluate all 5 designs as a whole, concluding in which of the concepts I will take forward to development and production.

The 'Mushroom' ottoman offers a new and different look to the ottoman concept. Positives from this design include the unknown nature of the material, wicker, aswell as the shape. However the ottoman is not easy to use. The heavy nature of the lid means allot of effort is used to lift the lid, and even with a mechanism aiding this, it is still a major safety risk. Also the wicker material is 'awkward' according to the client. Therefore I will not be developing this product.

The 'Traditional' ottoman unit has the potential to be a lovely piece of furniture. The drawer system makes previously unstorable objects storable in such a big unit. However the traditional shape and design does not conform to 'expressionist' design required by the client. However the swade covered lining to the storage elements is liked by the client. Due to this concept not being 'exciting' enough I will not be developing this unit further. However, I will take certain positive elements, liked by the client forward, such as the swade lining.

The 'Gamble' ottoman has a very pleasing shape and look, and really does look like I imagined. The curve at the back of the ottoman is smooth and elegant, with the contour of the inlay behind working well. The client likes the shape and materials used, however, the heavy nature of the unit is a draw back. However, my personal opinion is that this concept is already over developed, and development itself would not change the design. However, I will take forward successful elements from this concept, including the inlay idea and shape, to material research and shape development.

The 'Sunburst' concept is one of real design. The amount of exploitation in the shape of the ottoman is unseen in the other 4 ideas. The shelf idea also shows that the storage within an ottoman concept doesn't have to be a traditional deep 'box'. However the unit is not functional due to it's instability. The client likes the overall look and detail in the design, and I personally think this design has the most potential within development and therefore this design is the one I will take forward to development and production.

However, the final design idea has some points which I will consider in the development of the 'sunburst' concept. I like the inclusion of metal in the 'Ornate Metal' design. The strength of metal can be taken advantage of within furniture, and should be considered especially in the leg part of the ottoman. Also the scoring of a design into the lid could be used, and possibly an alternative for the inlaying of a design.

The decision to continue development into the 'Sunburst' has been considered after evaluating all other 4 designs. As discussed successful elements of the failed concepts will be taken forward to create a final product where all elements are successful in form and function.

Client Comment: ' I agree with the designer here. The 'Sunburst' design holds a serious amount of potential, once it has become a functional product. All other 4 concepts have good points which can help in the final designing process. However, the 'sunburst' ottoman shape and design has a long way before it is truly expressionist.'

Figure 4.6 *A student's review of a design idea against the specification criteria with client and third-party feedback*

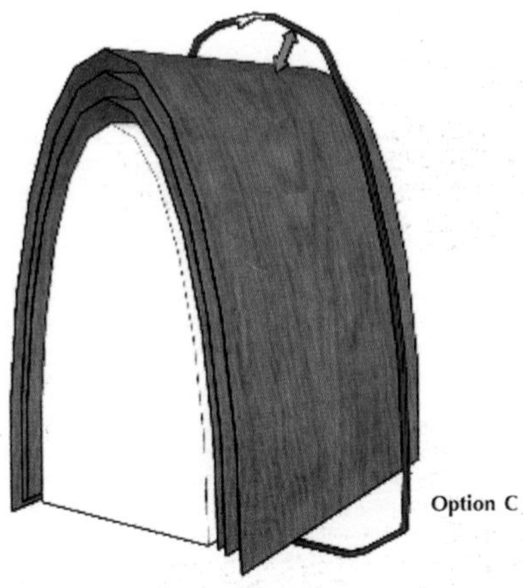

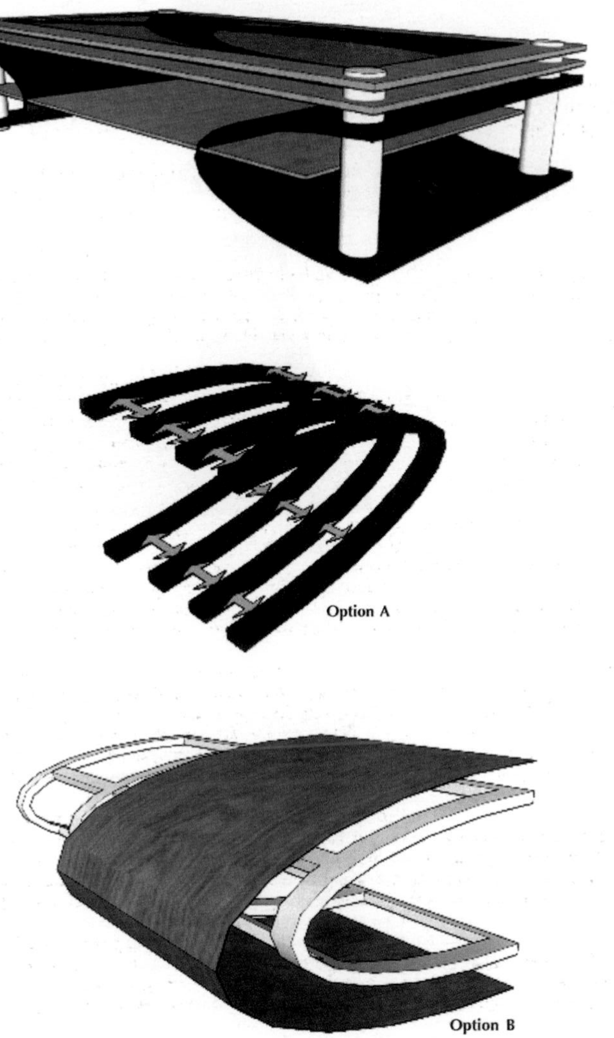

The construction process for the legs of the ottoman

The legs of the ottoman will mainly for fill a non structural role, as the top of the ottoman is supported by the metal supports. However, their shape and look is extremely important to the overall impression the ottoman gives. The curves have to be smooth, but solid to achieve this. Therefore I have chosen a selection of 3 possible construction options which I will consider for this part of the ottoman.

Option A is a method taken from the original design concept. This method creates the curve by using the CNC router to individually cut out curved sections which are then glued and dowel jointed together horizontally. The dowels have to be included in order for stability between the individual curves. The CNC router will give an accurate and unvarying size between the first and final curve, creating a smooth finish. However, this method has some draw backs. The shape, in this method, ie. individual curves, puts huge stresses on the turning point of the curve. Although the curve is not load bearing, if it does come under compression, it would most likely break under this method. Also, although a smooth finish would be produced, the joining between curves would be obvious and would harm the professional look of the product, which is important when considering the opinion the **client** expressed in the evaluation of ideas. He expressed that for the amount of money he is paying (£500), he wants "quality".

Option B is a frame method, which creates the curve using periodic supports across the width of the unit. The frame would be constructed out of simple "1 x 1" timber, while the ends, giving the shape, would be cut out on the CNC router. This method, eradicates the issue of stress at the apex of the curve, mainly due to the lack of sturdiness the method has. However, the smoothness and continuity across the curve is created by a single piece of veneer, placed ontop of the frame. However, this method also has draw backs. The frame will be hard to construct accurately, and the shape and design, would not hold much strength. The application of veneer over the frame, will create smoothness. However, due to the thinness of the veneer, the frame supports would be visible, as the veneer would become slightly tighter over them. Also the single veneer layer could be ripped accidentally.

Option C is a traditional, tried and tested laminating method. The curve will be constructed by moulding thin pieces of wood, including plywood and the American Black Walnut veneer around a mould. This mould will be constructed using the CNC router. Then, single layers of veneer and plywood will be applied, using PVA glue to form a laminated curve to the correct shape. To keep the shape while the glue is curing, rachett clamps will hold the layers down. This method produces a strong, and stable curve, which also looks good and smooth.

Client Feedback: "Option C seems the best option producing the best final product."

Construction Decision

Figure 4.7 An exploded drawing gives this student an opportunity to outline industrial and commercial manufacture when developing the product

A2 Level DTRM ~ Ottoman Project

The interchange between the lid and the top of the metal supports.

The original design did not take into consideration the problem highlighted in Figure A. The problem arises around the radius of the hinge being producing lateral movement, which would be obstructed by the metal supports of the ottoman. Therefore I have researched the following three options which I will consider below.

Option A is enlarging the holes around the metal supports which therefore allows for enough clearance for the lid to be lifted from it's base. The holes will be created on the 2D design and will easily be modified from the original plans. However the look of the ottoman I will think will be comprimised. The look of large holes within the wood, seems to look like a miss calculation. Almost like a mistake has happened, and the holes are to big. Therefore I do not think this is a good option as the product will look "silly" and shody.

Option B is slightly changing the design of lid so the problem does not arise. Instead of the ottoman having it's lid across the whole length of the unit, only part of the ottoman lid will lift up. The lifting section will be the middle part, which includes the inlay. This therefore means that the legs will not obstruct the lid, and therefore the top layer of Western Red Cedar can be fitted flush to the metal support. However, I think that this does not look very impressive, and also means the lid is significantly weakened because of less material surrounding the inlay design. It also does create problems for the contruction of the inlay. This lid method, would mean that the inlay could not be constructed by a jig saw method which I very much favour.

However, Option C seems to produce the best way round the problem. Option C uses a fixed plate ontop of the lid, giving the impression that the legs for the ottoman go throughout the product. This option allows for the lid of the ottoman to also be supported beneath by the metal supports. The plate on the lid itself can be created using the lathe, and will be finished to the same look as the supports below making the leg seem to come up through the top Cedar layer. This option allows for the original look to be comprimised the least. It allows for the form to be forfilled, without the function and form changing the original.

Client Comment: ' Option A, would look like a mistake had occured in construction. Option B is to much of a change from the original concept which I liked so much. However, Option C does allow for the look of the ottoman to remain the same, aswell as being a clever way around the problem.

Construction Decision

Details on the above model.

I produced this model in order to illustrate the problem which occur, and had to be overcome. The model is a cross-section of one of the back legs of the ottoman. The original idea was for the supports of the ottoman to be flush with the Western Red Cedar. This concept is shown by line A. However, as the model demonstrates, if the side of a hole where to be at line A, the lid would not open. However, line B, represents the tightest point the edge of the hole could be. With Line B being the edge of a hole, the lid would clear the supports.

This model demonstrated a problem with the original concept of the lid. From this, I chose 3 options which would overcome the problem, which have been evaluated on this page.

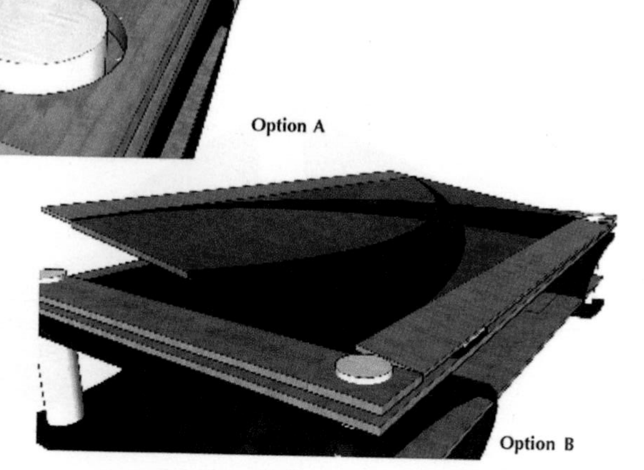

Figure 4.8 Part of a student's development of a product using modelling to test ideas

To be successful you will:

Assessment criteria C. Design and development: Develop

Level of response	Mark range
Produce a final design proposal through development that is significantly different and improved compared with any previous alternative design ideas. **(1 mark)** Present a final design proposal that includes technical details of materials and/or components, processes and techniques. **(1 mark)** Produce scale models using traditional materials or 2D and/or 3D computer simulations in order to test important aspects of the final design proposal against relevant design criteria. **(1 mark)** Use client/user group feedback for final modifications. **(1 mark)**	7–10
Develop appropriate designs that use details from alternative design ideas to change, refine and improve the final design proposal. **(1 mark)** Present a final design proposal that includes some details of materials and/or components, processes and techniques. **(1 mark)** Use modelling using traditional materials to test some aspects of the final design proposal against relevant design criteria. **(1 mark)**	4–6
Present development from alternative design ideas that are minor and cosmetic. **(1 mark)** Present a final design proposal that includes superficial details of materials and/or components, processes and techniques. **(1 mark)** Use simple models to test an aspect of the final design proposal against a design criterion. **(1 mark)**	1–3

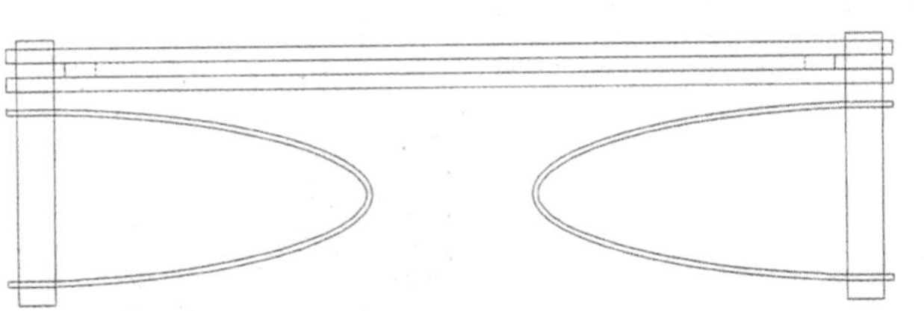

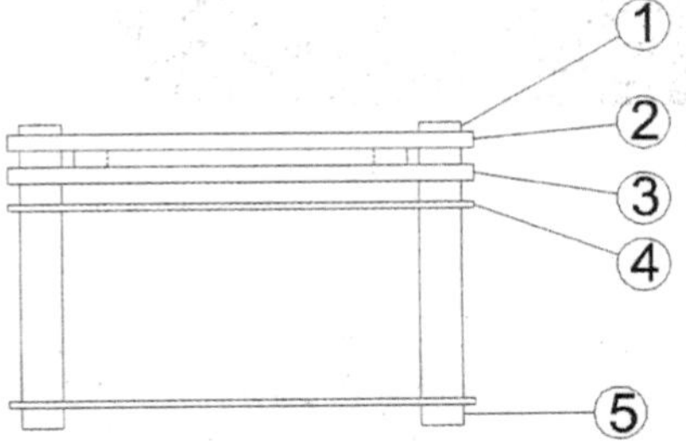

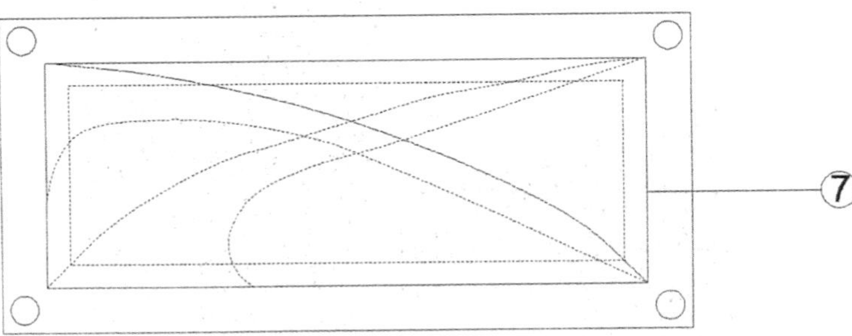

Drawn by: Richard Watton
Page: 1 of 4
Title: 3rd Angle Orthographic Projection
Do not Scale
All dimensions in mm

Figure 4.9 *CAD drawings by a student for an ottoman style storage unit*

Communicate (6 marks)

When presenting your design and development work, it is essential that your ideas are communicated effectively. This can be achieved in the following ways.

Through design and development work

You should show evidence of 'design thinking' using any form of effective communication that you feel is appropriate. However, you should try to use a range of skills that may include freehand sketching in 2D and 3D, cut and paste techniques and the use of ICT. It is important to demonstrate a high degree of graphical skill, which will be shown through the accuracy and precision of your work.

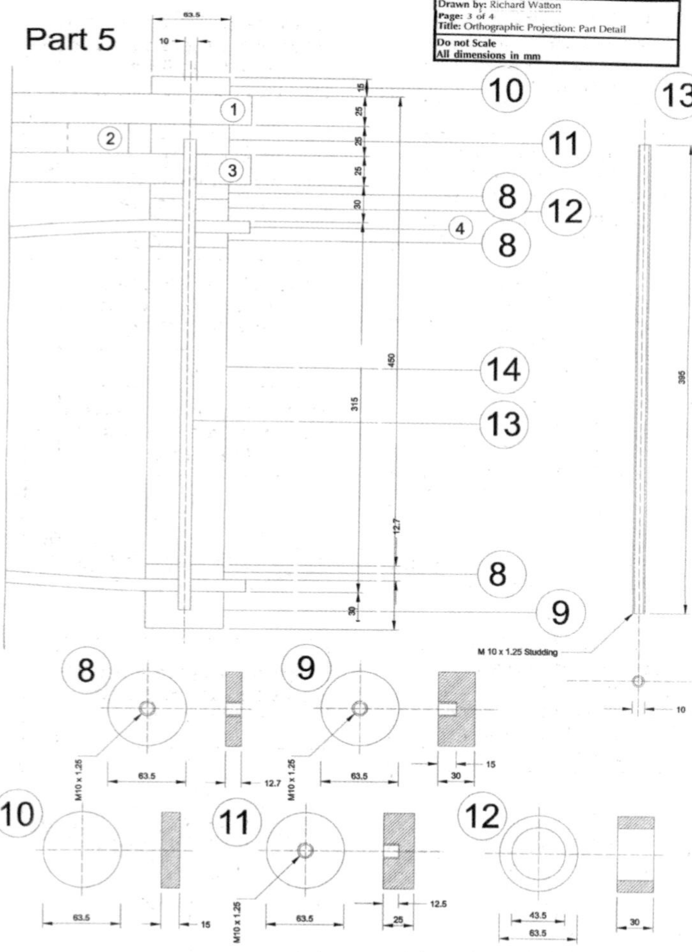

Figure 4.10 CAD drawing by a student showing specific detail dimensions to enable third-party manufacture

When using ICT, you should ensure that it is used appropriately rather than simply for show. For example, specialist computer-aided design (CAD) software to produce 3D rendered images is likely to be more appropriately used as part of development or final presentation, rather than for initial ideas.

Through presentation graphics and technical drawings

To effectively communicate final designs a range of skills and drawing techniques should be demonstrated, which could include:

- **pictorial drawings** – isometric, planometric (axonometric), oblique and perspective drawings to convey a 3D representation of the product
- **working drawings** – 1st or 3rd angle orthographic, exploded assembly and sectional drawings to convey technical information
- **computer generated** – pictorial and working drawings, renderings, etc. using specialist software.

Through the quality of written communication

Annotation should be used to explain design details and convey technical information. You should make sure that the information is presented in a logical order that is easily understood. Specialist technical vocabulary should be used consistently and with precision. Information presented in this section should enable your design thinking and manufacturing intentions to be clearly understood by others and allow third-party manufacture of the final design proposal.

To be successful you will:

Assessment criteria C. Design and development: Communicate

Level of response	Mark range
Use a range of communication techniques and media, including ICT and computer-aided design (CAD), **(1 mark)** with precision and accuracy **(1 mark)** to convey enough detailed and comprehensive information to enable a third party to manufacture the final design proposal. **(1 mark)**	4–6
Use a range of communication techniques, including ICT, **(1 mark)** with sufficient skill **(1 mark)** to convey an understanding of design and development intentions and construction details of the final design proposal. **(1 mark)**	1–3

D. Planning (6 marks)

In this section, you will produce a detailed production plan that explains the sequence of operations carried out during the commercial manufacture of your product. It should feature appropriate commercial practices and focus closely upon the identified scale of production. Therefore, workshop practices for the production of your scale model or working prototype should not be described.

A work order, or schedule, should be produced that illustrates the sequence of operations used during commercial manufacturing. This could be evidenced in the form of a flow chart. The work order should include the order of assembly of parts and components, featuring the necessary tools, equipment and processes to be used during manufacture in volumes higher than 'one-off' production (unless the designed product is specifically a one-off item).

An important part of planning is the use of time, so you must ensure that you consider realistic timescales and deadlines. Where Gantt or time charts are used, you must ensure that they are detailed, cover all aspects of manufacture and include achievable deadlines.

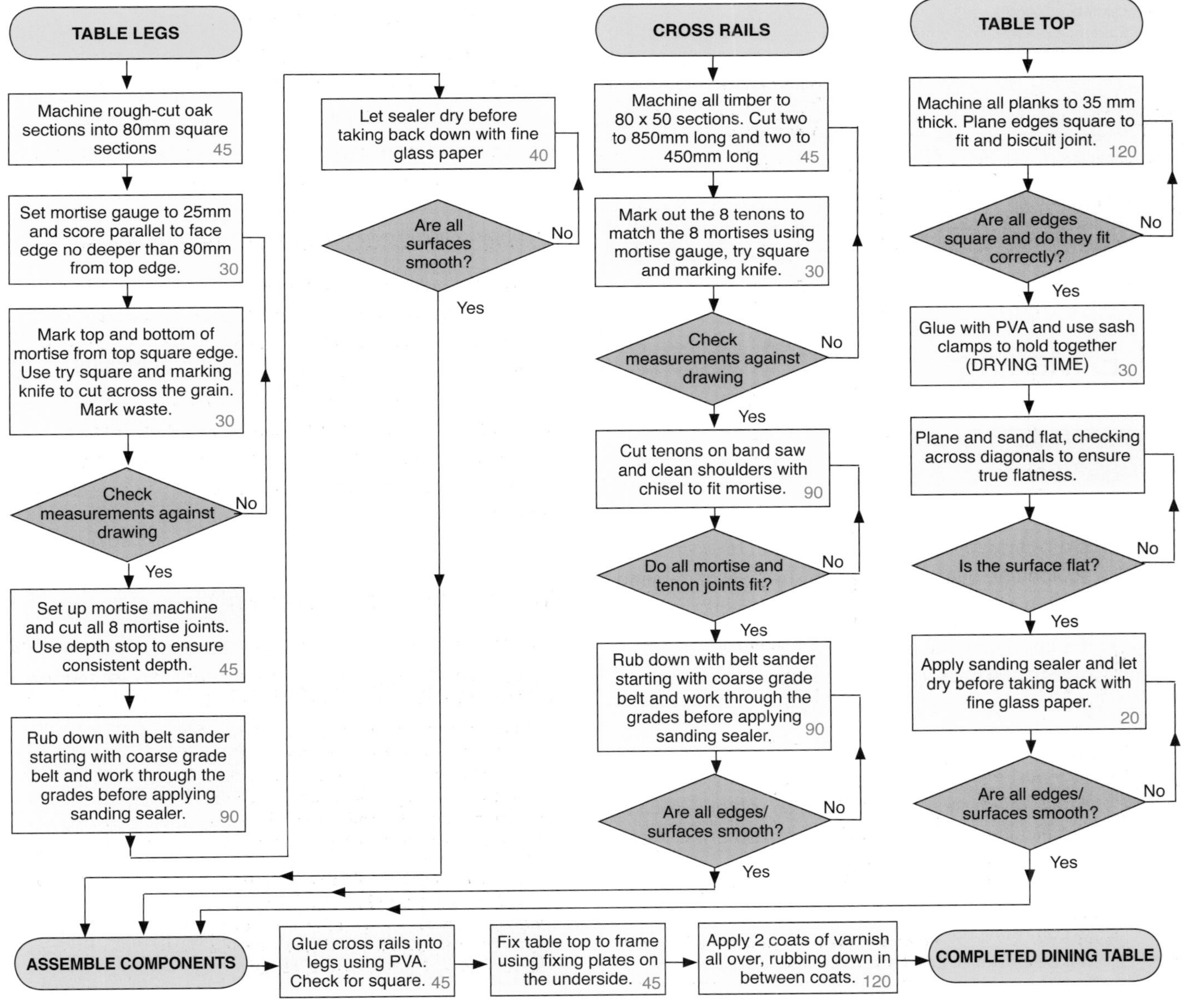

Figure 4.11 *A flow chart for planning manufacture of a dining table*

You should identify quality control (QC) points throughout the product's manufacture and describe all relevant quality checks used. This could be presented as part of a flow chart. Safety checks should also be included as part of planning.

To be successful you will:

Assessment criteria D. Planning

Level of response	Mark range
Produce a detailed production plan that considers the main stages of manufacture in the correct sequence appropriate to the scale of production. **(1 mark)** Evidence realistic and achievable timescales and deadlines for the scale of production. **(1 mark)** Show quality and safety checks that are justified. **(1 mark)**	4–6
Produce a production plan that considers the main stages of manufacture. **(1 mark)** Show reference to time and scale of production. **(1 mark)** Evidence superficial quality and safety checks. **(1 mark)**	1–3

E. Making

FACTFILE:

- The section *Making* carries the most marks and it is sub-divided into three areas: use of tools and equipment, quality and complexity/level of demand, making it easier for you to access the marks.
- It is important that all stages of the manufacturing process are photographed to evidence that the product is complete, expertly made, well finished, etc.
- You must ensure that photographs clearly show any details of advanced skills, technical content, levels of difficulty and complexity of construction, so that you can achieve the marks you deserve.
- It is unlikely that a single photograph will be enough to communicate all of the information required, so it will be better to take a series of photographs over a period of time during making.

Use of tools and equipment (9 marks)

You should demonstrate your ability to use tools and equipment with high levels of skill and accuracy and to select appropriate tools and equipment for specific purposes. It is important that you use a range of tools and equipment that allows you to fully demonstrate your skills.

Where computer-aided manufacture (CAM) is a feature of your work, you should make sure that there is plenty of opportunity within the product's manufacture to demonstrate other skills and competencies that you have acquired, for example, do not over-use computer-aided manufacture (CAM). Make sure that you make the majority of your product/model using hand tools and the appropriate machinery.

You should also work safely and be fully aware of the risks involved when using tools and equipment and of the precautions that should be taken to minimise those risks. Appropriate risk assessments for major practical activities could be recorded. Alternatively, stages where health and safety is important could be highlighted and explained within your photographic evidence of the manufacturing process.

Unit 2.4: Health and safety and **Unit 1: Making** for the procedures for carrying out a risk assessment according to the Health and Safety Executive (HSE).

To be successful you will:

Assessment criteria E. Making: Use of tools and equipment

Level of response	Mark range
Select tools and equipment for specific uses independently. **(1 mark)** Use tools and equipment with precision and accuracy. **(1 mark)** Show a high level of safety awareness, for self and others, when using specific tools and equipment. **(1 mark)**	7–9
Select appropriate tools and equipment with some guidance. **(1 mark)** Use tools and equipment with some skill and attention to detail. **(1 mark)** Show sufficient levels of safety awareness, for self and others, when using specific tools and equipment. **(1 mark)**	4–6
Select general tools and equipment with guidance. **(1 mark)** Use tools and equipment with limited skill and attention to detail. **(1 mark)** Show a limited level of safety awareness, for self and others, when using specific tools and equipment. **(1 mark)**	1–3

Vacuum-formed components continued . . . .

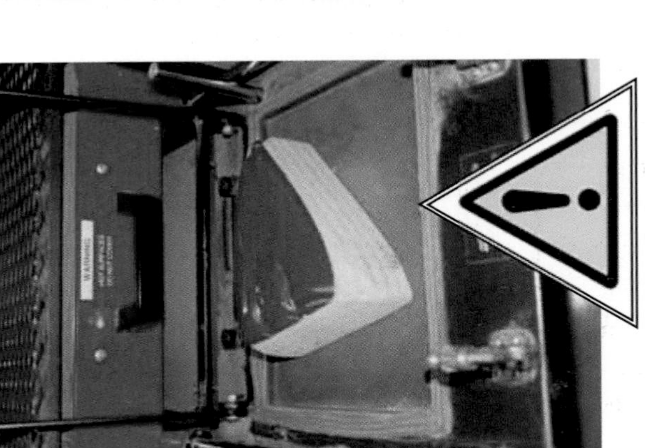

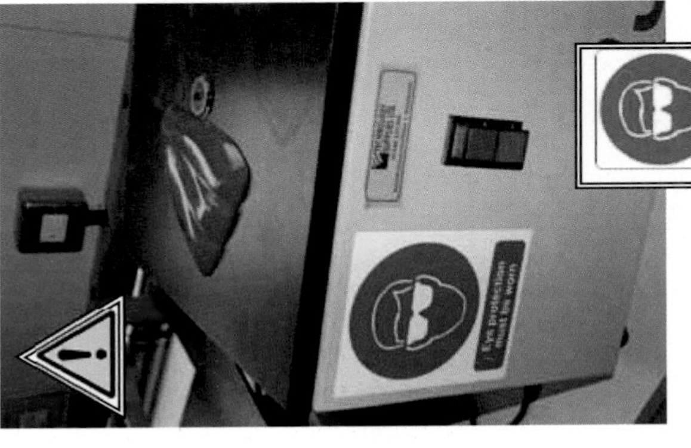

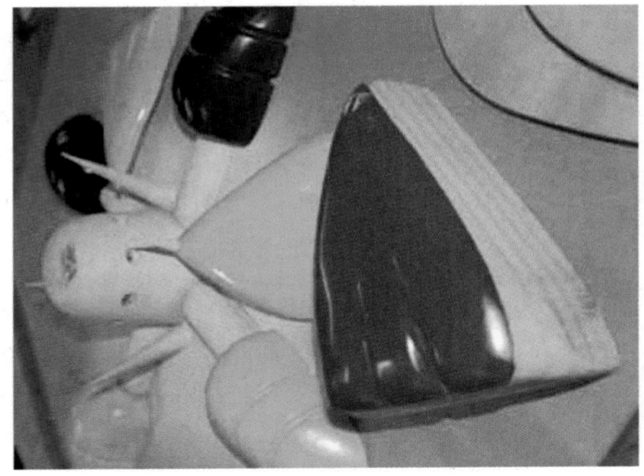

Selecting tools and processes

Due the success of vacuum forming the four pods (instead of turning two and sawing in half), I decided to use the vacuum former again for the sails.

After shaping the MDF mould, I used the vacuum former to produce FOUR identical copies of the sail components. It would have taken a lot more effort to produce these individually and I probably wouldn't have produced them all to the same standard.

Health & Safety

When using the vacuum forming machine I was extremely conscious of the heat it generated in order to soften the polystyrene sheet. Therefore, I was careful not to touch any hot surfaces.

The next hazard was the risk of trapping fingers in the many movable parts and clamping mechanisms. I made sure that I was paying full attention throughout the whole process.

Lastly, to prevent the plastic sheet from melting and burning I carefully timed the process and ensured that the machine was under my supervision for the duration.

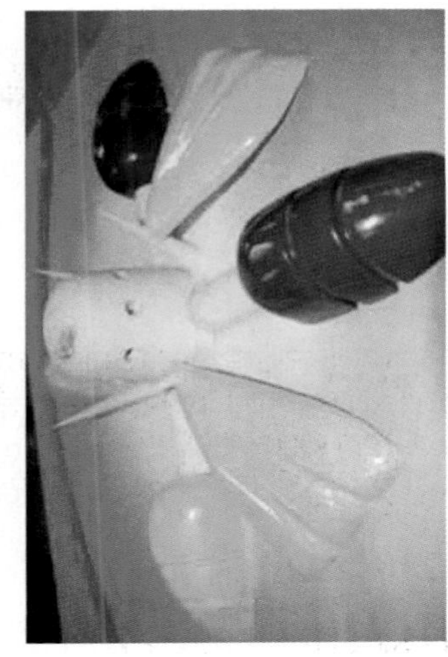

Model of theme park with vacuum-formed components

Figure 4.12 *Part of a student's documentation for selecting tools and identifying health and safety issues when making*

Quality (16 marks)

During the manufacture of your product, you should demonstrate your understanding of a range of materials and their working properties. In addition to this you should be able to select and justify the use of materials that are appropriate to the needs of the product and that match the requirements of your product specification. When selecting materials, you should be able to justify your choice by referring to material properties and suitability for their intended use. The selection and use of appropriate processes and techniques should enable you to produce a high-quality final product that fully matches the final design proposal in all respects.

It is important that all stages of the manufacturing process are photographed to evidence that the product is complete, expertly made, well finished, fully functioning, etc. For this reason you must ensure that photographs clearly show any details of advanced skills, technical content, levels of difficulty and complexity of construction, so that you can achieve the marks you deserve.

It is unlikely that a single photograph will be enough to communicate all of the information required, so it will be better to take a series of photographs over a period of time during making.

ACTIVITY:

At each stage in the manufacture of your product ensure that you have photographs of the components you are making. Use a digital camera, or a mobile phone, to take photographs that could then be easily downloaded onto your computer at school or at home to construct your pages.

Treat this activity like a diary of making the product where you write up what you have achieved in each lesson, problems encountered and explanations of why you tackled tasks in the manner you decided.

Do not leave this section until the end – you will have far more work to catch up on and you may well fail to document some important stages in the manufacture process.

To be successful you will:

Assessment criteria E. Making: Quality

Level of response	Mark range
Display a detailed understanding of the working properties of materials used **(1 mark)** with justification for your selection. **(1 mark)** Display a justified understanding of the use of manufacturing processes. **(1 mark)** Produce a high-quality product **(1 mark)** that matches all aspects of the final design proposal **(1 mark)** and is fully functional. **(1 mark)**	11–16
Display a good understanding of the working properties of materials used **(1 mark)** with relevant reasons for your selection. **(1 mark)** Display a good understanding of the use of relevant manufacturing processes. **(1 mark)** Produce a product that matches the final design proposal **(1 mark)** and functions adequately. **(1 mark)**	6–10
Display a limited understanding of the working properties of materials used **(1 mark)** with limited reasoning for your selection. **(1 mark)** Display a limited understanding of the use of manufacturing processes. **(1 mark)** Produce a product that barely matches the final design proposal **(1 mark)** and functions poorly. **(1 mark)**	1–5

Complexity/level of demand (9 marks)

It is important that you demonstrate demanding and high-level making skills in order to achieve high marks. For this reason it is very important that the manufacture of your product offers enough complexity and challenge to gain the maximum credit possible.

The level of complexity of the intended product will already have been established through the finalisation of the design proposal, so it is important that you consider this at an early stage to maximise your potential when manufacturing the product.

Try to set challenges and demands appropriate to your skill levels and beyond, so that you do not work within your 'comfort zone' and fail to achieve what you are actually capable of.

Avoid producing simplistic and undemanding work that, however well it is manufactured using appropriate tools, equipment and processes, is unchallenging. This approach cannot result in high levels of credit.

'Jsing the lines which I had marked out, I used the biscuit jointing nachine to cut out the biscuit joints. I ensured accuracy by making sure he machine was resting horizontally on the plank, and I ensured safety by earing safety glasses.

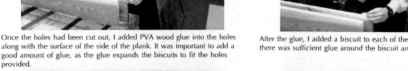

Once the holes had been cut out, I added PVA wood glue into the holes along with the surface of the side of the plank. It was important to add a good amount of glue, as the glue expands the biscuits to fit the holes provided.

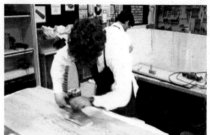

After the glue, I added a biscuit to each of the holes, and made sure that there was sufficient glue around the biscuit and the holes.

Above, after the sash clamps have been tightened, I am removing any excess glue with a damp cloth. The excess glue should be removed very quickly, as if left on the surface of the plank it will stain, and will result in the stain not penetrating the grain, and so leaving an ugly light patch.

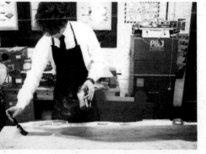

After running through the previous process with the other half of the 4 plank top, I can join the two half together. In addition, this time I had clamps over the top of the plank to reduce the chance of bowing or the plank not joining flat. After this I also removed the excess glue with a damp cloth, and the water marks can be seen on the surface of the plank.

Once the glue has set, I dampened the surface with a wet cloth. Despite, the stain being an oil based one, this rose microscopic fibbers up. Then I used an electric sander with a smooth 240 glass paper to remove these fibres. This makes the surface of the plank very smooth before staining. I also used this process to remove dents into the wood which had occurred due to knocks on this very soft wood.

bove I am applying a light English Oak finish to the Western Red Cedar. o do this, I am using a brush to evenly distribute the stain over the surface f the plank. It is important to keep the brush moving as if the stain gets me to dry, before the whole plank is covered, the different drying times ill result in lines and shades occurring over the surface.

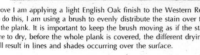

After the stain has been applied, I then applied a bees wax finish over the surface. After applying the wax, it is recommend you leave the wax for 15 minutes before using a non lint cloth to remove, and buff up the finish which I am pictured doing above.

Above I am spreading the contact adhesive over the back of the veneer which I will apply to once side of the MDF batton which will be glued onto the lower seat plank of western red cedar.

Realisation 3

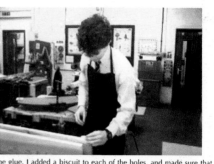

A2 Level DTRM ~ Ottoman Project

Figure 4.13 Part of the comprehensive documentation of the making process

Figure 4.14 An example of a product requiring a range of skills and attention to detail

To be successful you will:

Assessment criteria E. Making: Complexity/level of demand

Level of response	Mark range
Undertake a complex and challenging making task. **(1 mark)** Include a wide range of skills **(1 mark)** demonstrating precision and accuracy in their use. **(1 mark)**	7–9
Undertake a reasonably complex making task that offers some challenge. **(1 mark)** Include a range of skills, **(1 mark)** demonstrating attention to detail in their use. **(1 mark)**	4–6
Undertake a simple and undemanding making task. **(1 mark)** Include a limited range of skills **(1 mark)** requiring little attention to detail in their use. **(1 mark)**	1–3

F. Testing and evaluating (10 marks)

Once you have completed the manufacture of your product, you should carry out tests to check its fitness-for-purpose with reference to commercial techniques where possible.

Your finished product should be tested under realistic conditions, wherever possible, to decide on its success using the points of specification to check the product's performance and quality. You should describe in detail any tests carried out and justify them by stating what is being tested and why. Tests should be objective and many should be carried out by the client/user group. In addition, the involvement of other potential users would be a reliable way of gathering unbiased and reliable third-party feedback.

Well-annotated photographic evidence is a good tool to use when describing testing. You should use the results of your testing and views of the client/user group to help evaluate your final product.

Your evaluation should relate to the measurable points of your product specification and should be as objective as possible. Use the information from your testing, evaluation and client/user group feedback to make suggestions for possible modifications and future improvements to the product. Suggestions for modifications should focus on improving the performance of the product or its quality, and not simply cosmetic changes.

ACTIVITY:

Many students often overlook the testing and evaluation of a final product as 'not that important'. However, it is extremely important in wrapping up your project and bringing it to a logical conclusion. In order to carry out effective testing and evaluation of your final product you need to consider the following:

- Testing against your initial design specification to determine whether it satisfies the criteria. This will involve the justification of each specification point.
- Gaining third-party feedback from your client and/or user group using appropriate questionnaires or interview prompts.
- Writing an objective evaluation that discusses both the positive and negative aspects of your project.
- Suggesting modifications as a result of your evaluation. Remember that no design is ever perfect so there are bound to be aspects that need to be improved if you had more time. More importantly, there are bound to be technical problems that you did not have the time to address in detail that would need further development.

Finally, you should check the sustainability of your final product by carrying out a life-cycle assessment (LCA) to assess its impact on the environment. The most important life-cycle stages to consider when carrying out an LCA of your final product are below.

- **Raw materials** – What impact does the extraction of the raw materials in your product have upon the environment? Could you use fewer materials? Could you use recyclable materials?
- **Manufacture** – Are the processes you have identified energy efficient? Can manufacturing and assembly processes be simplified? How can the amount of waste produced be minimised?
- **Distribution** – How can transportation mileage be minimised? Can the design be simplified so as to reduce or lighten materials?
- **Use** – Will your product last a long time and can it be repaired if something goes wrong? How can you promote its efficient use? Can you use its green credentials to positively market your product?
- **End-of-life** – Can your product be recycled or reused? How can you reduce the amount of waste it produces from ending up as landfill?

LINKS TO:

Unit 3.4: Sustainability looks at these issues in greater detail.

To be successful you will:

Assessment criteria F. Testing and evaluating

Level of response	Mark range
Carry out a range of tests that are justified in order to check the performance and/or quality of the final product. **(1 mark)** Objectively evaluate, including third-party evaluation, considering most relevant, measurable specification points in detail. **(1 mark)** Suggest modifications that are justified from tests carried out; focus on improving performance and/or quality of the final product. **(1 mark)** Carry out relevant and useful LCA on the final product to check its sustainability. **(1 mark)**	7–10
Carry out a range of tests to check the performance and/or quality of the final product. **(1 mark)** Use objective evaluation with reference to most specification points. **(1 mark)** Suggest relevant modifications that are justified from tests that were carried out. **(1 mark)**	4–6
Carry out one or more simple tests to check the performance and/or quality of the final product. **(1 mark)** Use subjective and superficial evaluation with reference to a few specification points. **(1 mark)** Suggest only simple cosmetic modifications. **(1 mark)**	1–3

A2 Level DTRM ~ Ottoman Project

Modifications

After testing the final product, as well as evaluating against the original specification with the client, I have produced the following three modifications to the product which will improve the product further for it's created environment.

Modification 1

In response to the manufacturing and testing of the product, this modification would produce a stronger product and also would achieve the same look, but with less effort.

The modification is to use solid aluminium for the largest section, instead of the hollow tube which I used. Although I decided against this is the original material section of the project, due to cost, the extra cost would have been justified, as it would have produced a better product. The solid aluminium rod would have been turned down on the lathe, to have a 10 mm thread on one part, and 10 mm threaded hole on the other. This way the two parts could have screwed into the one another and would have reduced the need for welding the aluminium plates to the inside of the aluminium tube.

The welding of plates onto the inside of the aluminium tube is very tricky process as the plates have to be exactly horizontal and central inside of the hollow tube. This can be very time consuming, as there are so many components which have to be parallel and flat.

Client Comment I can understand why the designer wishes to make this modification. This would have allowed for a much more easy construction of the ottoman, and therefore would have allowed the manufacturer to spend more time on finishing the product, than production.

Modification 2

In response to the testing results and client's concern of the Western Red Cedar being damaged by corners of remote control and DVD boxes, within the storage ring, the following modification has been suggested to protect the cedar.

The modification is to place a layer of red Swede at the bottom of the storage ring, and so therefore over the cedar, protecting it. The colour of Swede is stronger and deeper than the cedar itself, but this complements the colours. The Swede would be glued to the cedar using contact adhesive, and the cut would be underneath the American Black Walnut storage ring. The addition of Swede would be a good modification to this soft wood

At present the soft wood is easily damaged. by the smaller living room objects. Unfortunately this means that marks are left on the surface of the wood, and therefore ruining the look of the wood.

Client Comment I like the idea of putting Swede inside storage area of the lid. However, the present situation of denting the wood is not really a problem. It just means that a certain amount of care has to be taken when placing, and removing objects from the storage area.

Modification 3

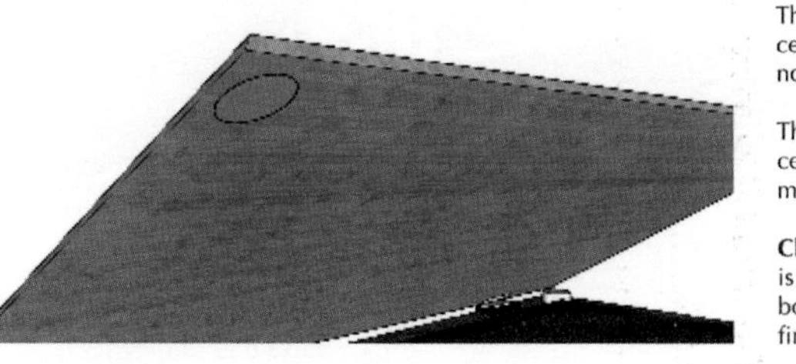

In response to the results from testing & the client's concern of the Western Red Cedar being damaged by the tops of the aluminium after constant use, this following modification protects the cedar from the aluminium tops.

The modification is to have circle of stained hard wood, around the area which the aluminium leg hits the western red cedar. The hard wood, could be oak, but it will be stained to a similar colour of the surrounding cedar. This hard wood, will not be damaged as much as the cedar by the aluminium top to the metal supports.

These circles of hard wood, would be located using the CNC router, and therefore accurately placed on the bottom of the cedar plank. This way, a collection of bruises and bash marks will not slowly gather around the areas where the cedar meets the aluminium legs.

Client Comment Once again I can understand why this modification is needed. However, like the 2nd modification, if care is taken when lowering the lid is being lowered, then the wood will not be damaged.However, the idea of a cap on the bottom of the lid will allow for less damage accumulating over time, keeping the piece of furniture to it's high standard of finish for longer.

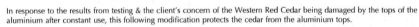

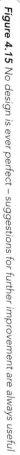

Figure 4.15 *No design is ever perfect – suggestions for further improvement are always useful*

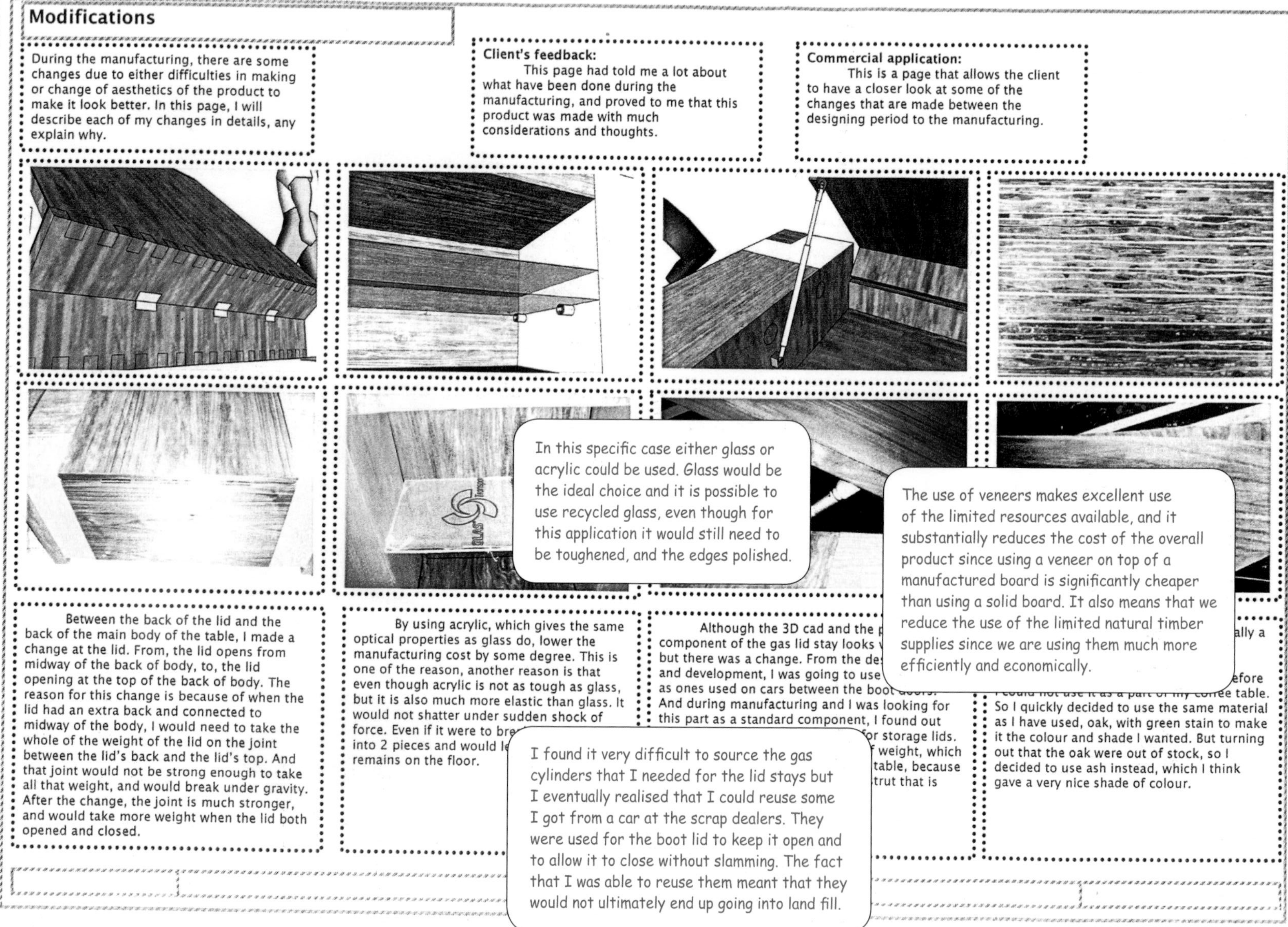

Modifications

During the manufacturing, there are some changes due to either difficulties in making or change of aesthetics of the product to make it look better. In this page, I will describe each of my changes in details, any explain why.

Client's feedback:
This page had told me a lot about what have been done during the manufacturing, and proved to me that this product was made with much considerations and thoughts.

Commercial application:
This is a page that allows the client to have a closer look at some of the changes that are made between the designing period to the manufacturing.

In this specific case either glass or acrylic could be used. Glass would be the ideal choice and it is possible to use recycled glass, even though for this application it would still need to be toughened, and the edges polished.

The use of veneers makes excellent use of the limited resources available, and it substantially reduces the cost of the overall product since using a veneer on top of a manufactured board is significantly cheaper than using a solid board. It also means that we reduce the use of the limited natural timber supplies since we are using them much more efficiently and economically.

Between the back of the lid and the back of the main body of the table, I made a change at the lid. From, the lid opens from midway of the back of body, to, the lid opening at the top of the back of body. The reason for this change is because of when the lid had an extra back and connected to midway of the body, I would need to take the whole of the weight of the lid on the joint between the lid's back and the lid's top. And that joint would not be strong enough to take all that weight, and would break under gravity. After the change, the joint is much stronger, and would take more weight when the lid both opened and closed.

By using acrylic, which gives the same optical properties as glass do, lower the manufacturing cost by some degree. This is one of the reason, another reason is that even though acrylic is not as tough as glass, but it is also much more elastic than glass. It would not shatter under sudden shock of force. Even if it were to bre[...] into 2 pieces and would le[...] remains on the floor.

Although the 3D cad and the p[...] component of the gas lid stay looks [...] but there was a change. From the de[...] and development, I was going to use [...] as ones used on cars between the boot [...] And during manufacturing and I was looking for this part as a standard component, I found out [...]or storage lids. [...] weight, which [...] table, because [...]trut that is

[...]ally a [...]efore [...] coffee table. So I quickly decided to use the same material as I have used, oak, with green stain to make it the colour and shade I wanted. But turning out that the oak were out of stock, so I decided to use ash instead, which I think gave a very nice shade of colour.

I found it very difficult to source the gas cylinders that I needed for the lid stays but I eventually realised that I could reuse some I got from a car at the scrap dealers. They were used for the boot lid to keep it open and to allow it to close without slamming. The fact that I was able to reuse them meant that they would not ultimately end up going into land fill.

Figure 4.16 *Environmental issues relating to the final design are highlighted*

Index